A RHETORIC OF SCIENCE

STUDIES IN RHETORIC/COMMUNICATION
Carroll C. Arnold, *Series Editor*

Richard B. Gregg
Symbolic Inducement and Knowing:
A Study in the Foundations of Rhetoric

Richard A. Cherwitz and James W. Hikins
Communication and Knowledge:
An Investigation in Rhetorical Epistemology

Herbert W. Simons and Aram A. Aghazarian, Editors
Form, Genre, and the Study of Political Discourse

Walter R. Fisher
Human Communication as Narration:
Toward a Philosophy of Reason, Value, and Action

David Payne
Coping with Failure:
The Therapeutic Uses of Rhetoric

David Bartine
Early English Reading Theory:
Origins of Current Debates

Craig R. Smith
Freedom of Expression and
Partisan Politics

A Rhetoric of Science:
Inventing Scientific Discourse

by Lawrence J. Prelli

University of South Carolina Press

Published in Columbia, South Carolina, by the
University of South Carolina Press

Manufactured in the United States of America

First Edition

Library of Congress Cataloging-in-Publication Data

Prelli, Lawrence J., 1955–
A rhetoric of science : inventing scientific discourse / by
Lawrence J. Prelli.
p. cm.—(Studies in rhetoric/communication)
Bibliography: p.
Includes index.
ISBN 0-87249-645-7
1. Communication in science—Philosophy. 2. Rhetoric—Philosophy.
3. Invention (Rhetoric) I. Title. II. Series.
Q223.P74 1989
500—dc20 89-16442
CIP

For My Teachers

CONTENTS

EDITOR'S FOREWORD

This is a book about the "topical logic" of scientific rhetoric. That there are rhetorical dimensions to doing science has been convincingly argued during the last four decades. What is the *conceptual* character of the rhetoric of science? is a question that has to date been treated in only piecemeal ways. In *A Rhetoric of Science* Lawrence Prelli argues that specially significant ideational qualities that are valued by all scientific communities constitute *topics* identifying the logical standards that purporting scientific discourse is expected to meet. From analyzing a variety of samples of scientific argument, Prelli concludes that the values embedded in the procedures called "scientific method" identify the topical-logical qualities scientific discourse is expected to display and by which the *logical* legitimacy of scientific claims are judged.

The concept of "topical logic" requires preliminary explanation because, although the notion and practice of such logic have a very long history, moderns have thought little about topical reasoning until very recent times. The important question that lacks systematic answer is: If scientists cannot justify their scientific claims in positivistic terms, how then are those claims "proved" to be scientifically reasonable? Prelli addresses this question directly, contending that claims are judged for their logical conformity to recurring criteria agreed upon as characteristic of "true" scientific thinking. He contends that scientific discourse is created and is evaluated for its consonance with this finite group of values that specify the logical requirements of "doing science" through discourse. Because those requirements are informally, but still logically, adopted and applied to scientific claims, they collectively identify the logical structure against which rhetoric aspiring to the status of being "scientific" is measured. As Mr. Prelli grants, other logical criteria can, of course, also operate in doing science. For example, the narrational logic that Walter R. Fisher describes in *Human Communication as Narration,* another volume in our Rhetoric/Communication Series, presumably operates in science as well as elsewhere. Sometimes principles of formal logic also operate. What Prelli offers is not, therefore, *the* rhetorical logic of science but *a* rhetorical logic—and a fundamental one—that is always used when discourse is evaluated for its scientific merits.

Understanding Prelli's discussion fully requires an understanding of the theory of rhetorical *invention,* as posited in ancient and modern rhetorical theory. Scholars well versed in the history and postulates of that theory will find little that is new to them in chapters 2 through 5. They may wish to concentrate their attention on chapter 1 and chapters 6 through 11. Mr. Prelli and his editor hope, however, that chapters 2 through 5 will sufficiently equip readers who are unacquainted with that theory to follow Prelli's later applications of those constructs to scientific argumentation.

The University of South Carolina Press welcomes this book to its Rhetoric/Communication Series, believing that Professor Prelli's work sheds fresh light on how scientific claims are legitimated. The book reveals the *rhetorical logic* that is never irrelevant in proposing and defending scientific claims. This logic has never before been examined systematically, although several recent writers have touched on the system without perceiving its peculiar relation to theory of rhetoric and its difference from systems of formal logic. We hope, too, that this book will show readers how they can uncover the peculiarities of the special topical logics used in scientific specialties and in domains that do not purport to make "scientific" claims.

Carroll C. Arnold
Editor, Rhetoric/Communication
University of South Carolina Press

ACKNOWLEDGMENTS

No book is the product of its author's efforts alone. Colleagues, students, and friends contribute through critical readings and reactions, stimulating conversations, encouragement, and less tangible forms of support. Among those to whom I am indebted, special thanks go to Carroll C. Arnold. His editorial guidance improved significantly the substance, structure, and style of this book. I learned much about rhetorical studies and the world of scholarship from our interactions. Wilburn Sims read several drafts of the manuscript, helping immeasurably with reactions, corrections, and friendly encouragement. Spirited late-night conversations with Lawrence H. Stern helped me understand more incisively the social history and technical intricacies of the memory-transfer controversy. John Lyne's critical commentaries contributed to making this a better book. Other colleagues and friends who stimulated thought and thus helped to improve portions of the book include John Adams, Marcy Dorfman, Nancy Dunbar, Val Dusek, Roger Pace, and Larry Underberg.

I am indebted to many other people for creating the kind of research environment needed to complete this book. I am grateful to the University of New Hampshire's Humanities Center and Graduate School for providing summer faculty fellowships, and to the Liberal Arts College at the same institution for a faculty research support grant. Dean Stuart Palmer and Dean Raymond L. Erickson provided additional resources to help meet obstacles encountered while working toward completion of the manuscript. The research and interlibrary loan staff at the University's Dimond Library always provided cordial assistance when needed. I appreciate Deborah Whitmoyer's painstaking care in typing preliminary drafts of the manuscript. Special thanks go to Terri Winters, whose word processing wizardry helped to produce the final manuscript, and who offered other forms of support.

A RHETORIC OF SCIENCE

1

INTRODUCTION

The idea that there is a "rhetoric of science" may strike some as peculiar. "Rhetoric" and "science" are terms that evoke conflicting conventional understandings. In conventional wisdom "rhetoric" is, at best, thought of as the rules of styling and organizing discourse about substantive matters that have no necessary connection with "rhetoric." A cynical view conceives of rhetoric as junk food for the mind, stimulating the senses with the spices and sweets of emotion, leading ultimately to intellectual malnutrition. Science, on the other hand, is thought by many to be the pinnacle of human achievement. Conceptualized as a body of stable, certifiably "objective facts" interrelated validly by rigorous logic, science is seen as promising progress in knowledge and in human control over the phenomena and forces of the universe.

In this book I shall offer fundamentally different views of rhetoric and of science. I intend to inquire into how the arguments that comprise scientific communication are designed and on what grounds they are and ought to be weighed as scientific claims. In doing so I propose to go a step beyond the now numerous contentions by scientists and others that scientific discourse has rhetorical dimensions. My central interest is in the rhetorical dimensions of *creating* and *evaluating* scientific communication.

Scientists and scholarly commentators on science have become increasingly aware that apodictic proofs are rare, even in science, and that concepts associated with formal logic are insufficient to describe the activities of "doing science." They have therefore undertaken to reconsider how to talk and think about scientific work. I shall argue that when scientists address claims discursively to other scientists, they make a special kind of *rhetoric*. I shall undertake to elaborate the basic features of the discursive activities scientists engage in and, by focusing on how they *invent* scientific discourse, offer a theory of rhetoric that is specific to scientific activity. These undertakings are parallel to but not identical with philosophical, historical, and sociological studies of science.

Ernan McMullin reflected the contemporary state of philosophy of science when he came to the conclusion that the "rationality" of science is being threatened. In his view, although the work of Feyerabend, Kuhn, Polanyi, Toulmin, and others has dismantled the foundationalist program of the Vienna Circle, the reaction to logicism and foundationalism has gone so far that the objectivity and rationality of science needs to be defended. However, McMullin argued that the logical positivists erred in restricting scientific rationality to reasoning based on formally logical criteria. According to this view,

> The formal relations between evidence and hypothesis are what primarily constitute the "logicality" of science. They can be made at least partially explicit as a set of formal rules of procedure, a logic which helps determine the validity (or in a weaker sense, the plausibility) of a proposed hypothesis-evidence linkage.[1]

Positivists sought, among other things, to distinguish science from purportedly less reliable modes of human inquiry on the basis of its allegedly formal logicality. This, McMullin claimed, is but one "face" of science—the logical one.[2]

The logical positivists' stress on the formal logicality of scientific rationality left their program open to extensive critique from a number of quarters. Popper criticized the methodological emphasis on verification and induction.[3] Polanyi and Hanson argued that the distinction between theoretical and observational terms was untenable since all observational reports depend on the conceptual schemes and expectations of the observers.[4] The historian of science, Thomas Kuhn, argued that the criteria of formal logicality, with its rules of validation, apply only in those periods of "normal science," which are characterized by consensual agreement on a "paradigm" for doing science.[5] In periods of "crisis," science undergoes revolutionary transformations wherein old scientific commitments decline as new ones are forged, and for reasons other than those specified by formally logical criteria.

Toulmin shared Kuhn's belief that the history of science supplied the key for discovering the characteristic patterns of scientific rationality.[6] Unlike Kuhn, he contended that science does involve a steady, evolutionary transformation of concepts. As concepts shift and undergo variation, there can, however, be no foundational propositions. For Toulmin, rationality is therefore not formal logicality; it is conceptual appraisal and selection for purposes of achieving "richer modes of understanding" when new and unforeseen experiences are confronted.

Feyerabend contended that comparative empirical testing cannot distinguish which theory is "best" because theories prescribe mutually incompatible ways of looking at the world.[7] No theory can account fully for all of the facts. This view directly challenged the core assumption supporting positivism: that theories can be placed in formal comparison with one another on the basis of which one theory can be preferred over another.

Popper, Polanyi, Hanson, Kuhn, Toulmin, and Feyerabend differ in many respects on how to characterize science correctly. I have cited a few of their topics in order to make a general point: their various positions, taken collectively, make clear that science has an other-than-formally-logical "face." McMullin has called it an *interpretive* face:

> It could be called *interpretation;* or perhaps *patterning* would also do. It is the attaching of meaning, the perceiving of structure, the recognition of something as an instance. These are not quite the same, and for some purposes it might be important to distinguish them. But they can be meaningfully grouped together as a loose unit when one contrasts them with the logicality aspect of science. Where logical inference is discursive, interpretation is immediate, intuitive. Where inference is explicit, interpretation tends to be implicit, that is, it cannot be reduced to explicit criteria. Where inference is formal, interpretation is material, dependent upon context, incapable of being expressed in a single abstract rule. Where inference is mechanizable, interpretation involves personal skills that do not lend themselves to mechanical simulation. Where inference is non-controversial, interpretation can be very controversial indeed.[8]

I believe with McMullin that there is in addition to the formally logical face of science an interpretive one that presents nonlogical but not irrational features. However, what McMullin could not see was that if science has an interpretive face, doing science will necessarily have *rhetorical* aspects. I shall show further that there is even a "logical face" of scientific practice that involves a scientist in *rhetorical* choosing based on criteria other than those of formal validity, but I defer developing that somewhat technical point until I have presented a full theory of rhetorical invention.

The logical face of science is characterized by a high degree of agreement and consensus concerning scientific knowledge. Logical positivists located the agency of consensus in the technical logic of scientific methodology. Traditional history and sociology of science complement this view. For instance, historians presented the history of science as evolving progressively toward universal consensus about positive knowledge through use of scientific methods to distinguish truth from falsehood. Thus, old ideas and practices were "progressive" if they approximated the thought and methods which are consensually accepted within present scientific fields, and they were erroneous or blocked progress if they did not.[9] Sociologists sought to explain processes of consensus formation in science by examining social influences and constraints on scientists rather than the technical logic of science. For instance, much attention was directed toward understanding the normative structures governing social activities among scientists, including how reward systems sanction "normal" and discourage "deviant" behaviors.[10] This was the sociological complement to the technical logic formulated by the philosophers; it locates normative prescriptions that promote maximally efficient usage of scientific methods and the dissemination of their results.

The interpretive face of science is characterized by controversy, disagreement, dissensus. Scholarly commentators, perceiving that the logic of scientific method did not exhaust processes of scientific knowing, probed the psychological dimensions of scientific creativity, conflict, and change. Historians of science challenged received views about the progressive evolution of scientific knowing by showing the history of science as marked by periodic clashes between incompatible world views.[11] Sociologists of scientific knowledge have focused attention directly on technical or cognitive conflicts about substantive scientific matters, arguing that scientific knowledge is not "found" or "discovered" by applying formal algorithms which compel agreement and secure consensus, but is itself socially constructed and negotiated.[12] The resulting state of affairs in general study of science has been summarized acutely by Laudan:

> Students of the development of science, whether sociologists or philosophers, have alternately been preoccupied with the explanation of consensus in science or with the analysis of disagreement and divergence. These contrasting foci would be harmless if they represented nothing more than differences of emphasis or interest. After all, no one can fix simultaneously on all sides of any question. What creates the tension is that neither approach seems to have the explanatory resources for dealing with both phenomena. More specifically, whatever successes can be claimed by each of these models in explaining its preferred problems are largely negated by its inability to grapple with the main problems of its rivals. Thus, earlier sociological and philosophical models of science adopt such

> strong assumptions about the consensus-forming mechanisms which they postulate to explain agreement that it is difficult to make much sense of the range and character of scientific disagreements and controversies. The more recent models leave us largely in the dark about how scientists could ever reasonably resolve their differences in the definitive fashion in which they often do terminate controversies. We need to be able to explain both these striking features about science; existing accounts lack the explanatory resources to tackle these two puzzles in tandem. Many of the fashionable recent approaches are at least as deficient as those they would replace.[13]

Some of the major problems Laudan points to exist because none of the standard disciplinary approaches to the study of science focuses systematically on science as a special kind of discursive activity. As a number of scientists and rhetoricians have shown, "doing science" inevitably entails making arguments that are informal, material, contextual, and controversial.[14] The efficacy of scientific discourse is constrained by contextual features that cut across logical and psychological, technical and social, cognitive and moral influences on scientific activity. The *creative* processes by which such discourse is conceived and framed have not been intensively examined as communicative, that is *rhetorical,* activity. Neither has there been intensive examination of how rhetorical choices *legitimate* scientific claims.

THE IMPORTANCE OF RHETORICAL INVENTION

To inquire into the rhetorical features of doing science entails exploring the informal, logical principles and practices that are applied in creating and evaluating scientific arguments. Among other things, one must ask what special kind of criteria govern the practice of scientific argumentation if they are not strictly formal. This brings one face to face with what Stephen Toulmin called the "general problem for comparative applied logic": determining "what, in any particular field of argument, the highest relevant standards will be."[15] One way of identifying the "highest relevant standards" in any field of argument is to inspect the valuational principles the pertinent community of thinkers expects discourse to meet. Accordingly, in Part II of what follows I will show that arguers in science continuously use an identifiable, finite set of value-laden topics as they produce and evaluate claims and counterclaims involving community problems and concerns. Some who have written on the rhetorical dimensions of science would disagree, but I insist that these dimensions include the applied principles of practical, rhetorical logic.

As this is being written, there exists but one other book devoted explicitly to investigating the comparative roles of rhetoric and logic in establishing scientific claims.[16] In that work, as elsewhere, Maurice A. Finocchiaro distinguishes *argument* from *rhetoric.* He sees rhetoric as essential in scientific discourse but subordinate to logic. This distinction prompted one reviewer to claim that Finocchiaro's book is more properly subtitled the "*Logical* Foundations of Rhetoric and Scientific Method," rather than the "*Rhetorical* Foundations of Logic and Scientific Method."[17] I make the inclusive, synthesizing claim that *there is a rhetorical logic in science.* In what follows I will articulate a conception of rhetoric that comprehensively informs the processes involved when scientists make and evaluate discourse as science. This will require (1) providing a clear exposition of what rhetorical action is and how it comes about, and (2) propounding a rhetoric of science that is technically sound as a theory of rhetoric and is representative of scientists' practices when they engage in discursive activity.

A caveat is needed here. The "rhetoric of science" I propose to offer is not science per se, but it must *reflect* the "logic" of scientific thinking—indeed, of scientific method itself. Neither is a "rhetoric of science" a philosophy of science, a history of science, a sociology of science, or a psychology of science; but, again, it must *reflect* the operations of philosophy, history, sociology, and psychology insofar as they constrain discursive practices involved in the doing of science. A comprehensive rhetoric is a theory of systematic, communicative practice; it posits a particular way of thinking about ideas and their communication. Greco-Roman culture possessed such a theory. That theory has been broadened and deepened by a number of modern scholars. As inherited, the general theory of rhetoric is attuned principally to creation of political, legal, and celebratory discourse. In the chapters to follow, I outline the basic tenets of this theory, particularly as they apply to the processes of *creating* discourse. From this general rhetorical theory, I then propose a specific theory based on the practices of scientists, to explain what scientists *do* when they engage in the many discursive activities that are parts of doing science.

As I use the term, "rhetoric" is the use of symbols to induce cooperative acts and attitudes. This is the historic understanding of what rhetoric is as action. A rhetor attempts to induce the participation of an audience in his or her particular orientation and its concomitant claims. Audiences have autonomy in choosing what they will and will not adhere to. Such autonomy is a central postulate of ancient and modern rhetorical theory, and contrasts with that which underlies Kuhn's psychological concept of conversion (which he calls "neural reprogramming"). Kuhn's conception implies what James called automatism—change that results from subliminal experience, activity, or "mental work."[18] Without denying that neural reprogramming can happen, rhetorical theory posits (as did James) that choosing among competing claims is also commonplace. Walter B. Weimer is a social scientist who has developed the philosophical underpinnings of a conception of scientific choosing. He defines scientific discourse as the making of warranted assertions about contingent claims that are based on "good reasons." As Weimer fully recognized, to say this is to treat scientific discursive activity as an attempt at achieving a *rhetorical transaction.*[19]

Viewing scientific discourse as based on "good reasons" is consistent with the postulates of inherited rhetorical theory, and it avoids the strains and struggles of accounting for scientific activity as fundamentally either rational or irrational, objective or subjective. It also is consistent with what actually happens in scientific exchanges. Audiences of scientists judge scientific claims, not with reference to the canons of formal logic, but against received community problems, values, expectations, and interests. The judgmental standards are located within situated audiences' frames of reference, not in logical rules that transcend specific situations for scientific claiming.

Scientific discourse is accepted or rejected on grounds of its *reasonableness*—given the issue at stake, the knowledge conditions of the scientific community, and the perceived expertise of the makers of the claims. Later in this book I shall elaborate this kind of claiming and judging under the rubric of "rhetorical reasonableness." It is sufficient at this point to observe that scientific communities exist by virtue of the same presuppositions about choosing among claims as are incorporated within standard rhetorical theory.

Inherited rhetorical theory posits that rhetorical creativity is systematic. Such theory identifies heuristics for constructing discourse. It offers principles and procedures by which rhetors can (1) identify the kinds of discursive choices they need to make, (2) discover the available options from which to choose in a given case, and (3) discriminate among themes and materials to locate those that will make claims maximally reasonable and persuasive given the kinds of judgments a particular audience is likely to form. The inventional aspects of rhetorical theory are oriented to practice. A rhetoric of science should therefore identify constructs and categories that derive from actual discursive practice in science, and should provide formulations that explain both how such discourse is made and how it is judged as *science*. I take the position that "the art of rhetoric," with its principles and methods of invention, can in fact usefully identify the kinds of discursive decisions scientists must make when they seek to render their claims reasonable in the eyes of scientific audiences.

I do not intend to argue that everything in science is rhetorical, without remainder. The issue is whether inspecting the principles according to which scientific discourse is created and judged can reveal significant aspects of scientific endeavors that might otherwise remain concealed. I believe that the perspective supplied by rhetorical theory can indeed bring to attention significant but largely unnoticed dimensions of "doing science."

In what follows I shall outline the leading features of rhetorical invention as presented in general rhetorical theory, then I shall derive from these principles a special account of how scientific discourse is created, how it is evaluated as rhetoric, and how it persuades.

Part I
RHETORICAL INVENTION

2

THE NATURE OF RHETORIC

There are at least as many conceptions of rhetoric as there are philosophical perspectives on human communication.[1] There are likely to be family resemblances among rhetorical theories sharing similar philosophical assumptions, but the content of rhetorical theories varies when we move among the philosophical perspectives of the sophists, Isocrates, Plato, Aristotle, Cicero, Quintilian, Augustine, Ramus, Vico, Campbell, Blair, Whately, Kenneth Burke, Perelman, and Weaver—to name only some of the more prominent theorists. Can we nevertheless extract some theoretical, common grounds from among the many answers to the question, What is rhetoric?

Duhamel contended that rhetoric—any rhetoric—pursues the object of teaching effective expression.[2] Regardless of philosophical and practical differences among particular perspectives on rhetoric, they share common ground in the idea that rhetoric deals with effective expression. This idea points to other generalizable features that make any specific perspective a rhetorical perspective.

There are five general features that inhere in the idea of rhetoric as effective expression. These features reflect the kinds of assumptions made when anyone thinks systematically and coherently about rhetoric. The broad classes of assumptions designate issues that are fundamental to thinking about effective expression itself. Thus, any concept of effective expression will reflect assumptions about (1) the *role of language,* (2) the nature of *audiences* and (3) *situations,* (4) the *criteria* by which materials for expression are evaluated, and (5) the *methods* of finding these materials in composing rhetoric.

Rhetorical theorists must engage the difficult question of what role words, language, or symbols have in effective expression. Language can be considered instrumental, as a means of communication. According to this view, we use words as pointers to clarify or evoke thoughts that correspond to our ideas and our impressions of things in the world. Frequently, however, these verbal means are viewed with suspicion. Empirically minded thinkers, those concerned with faithful expression of objective patterns among things in the world, view language that is not concrete, specific, and unambiguous as detrimental to effective communication.[3] On the other hand, rationalist thinkers who locate the source of meaning in stable and clear ideas are likely to view language as a problematic means of sharing ideas with others.[4]

Visions of the possible role for language in achieving effective expression are not limited to the instrumentalist stances of rationalists and empiricists. Other theorists view language as a fundamental, symbolic medium that mediates all meaning, whether of things in the world or thoughts in the human mind. As Wayne Booth explains, from this view language is "the medium in which selves grow, the social invention through which we make each other and the structures that are our world, the shared product of our efforts to cope with experience."[5] The fifth-century B.C. sophists and rhetoricians expressed similar views when they extolled the persuasive power of *logos* or speech, through which they thought reality was constituted. Leff summarizes the opinion of the great sophist Gorgias on this point: "We know what we know on the basis of encounters with actual situations as they are defined by the perceptual screen of language."[6] For sophists like Gorgias, there is no separate "reality" apart from what people are induced to believe through the persuasive power of language.

Whatever the role of language may be, a coherent perspective on rhetoric must also consider the nature of those to whom language is addressed. A concept of rhetoric must consider the audience. Is an "audience" an actual group or an ideal or universal construct? How much freedom do speakers and writers have in choosing their audiences? How are speakers and writers related to audiences—as superior (leader, enlightened authority, adviser) to subordinates (followers, ignorant masses, advisees); as equals, or psychically, like me internally addressing myself as an audience? More mundanely one might ask: What is the makeup of the audience? Is there homogeneity or heterogeneity regarding matters such as age, sex, education, religion, occupational level, social and political attitudes, beliefs and values, social class, culture? And rhetors might ask: What kinds of attitudinal, psychological, or behavioral responses from audiences are sought by speakers or writers? Theorists may add to this list of questions or disagree about the comparative importance of each question, but formulating an idea of audience must be a central feature of any understanding of rhetoric as effective expression.

Any perspective on rhetoric must also take into account the influence of context, occasion, situation. Again, this implies a series of questions:

What differentiates a rhetorical situation from other kinds of contexts? What kinds of rhetorical situations are there? Are they public, private, or both? Are they limited, say, to politics, jurisprudence, and the pulpit? Do rhetorical situations include contexts peculiar to theoretical or scientific exchanges through discourse? Are rhetorical situations universal or particular? Ideal or actual? Specific perspectives on rhetoric will view differently what constitutes rhetorical situations, but some conception of rhetorical situation is implied in all thought about effective expression. I shall have occasion later to point out difficulties encountered by theories of rhetoric that minimize the importance of situations.

Any perspective on rhetoric must also incorporate some notions about how the materials, the contents, of expression are evaluated as plausible, credible, or appropriate. We here confront the troublesome issue of how rhetorically made claims are legitimated. Many criteria for judging expressive content have been recommended: fidelity to authoritative religious, historical, philosophical, literary, or legal texts; correspondence with sensory experience; coherence and distinctness of valid reasoning; commensurability with ethical standards; appropriateness to communal norms and shared opinions. A conception of rhetoric inevitably includes, whether as an explicit or implicit assumption, some standards for determining the logical legitimacy and general suitability of materials for expression.

The final general feature of any conception of effective expression is some account of method in investigating or discovering materials to be expressed. The methods historically identified for this purpose range across classical dialectics, philology and hermeneutics, logic, empirical observation and experimentation, Hegelian and Marxist dialectics, and specifically and specially rhetorical inventional procedures.

These are five general features common to all conceptions of rhetoric. Given these features, I wish now to outline a comprehensive idea of rhetoric that is applicable to all actions and activities deemed suasory uses of symbols.

RHETORIC IS THE SUASORY USE OF SYMBOLS

My conception of rhetoric rests on the understanding that symbols—particularly linguistic symbols—are not tools or instruments that are manipulated to express separate ideational meanings with efficacy; symbols are the media through which all meaning is constituted. Following Burke, I take rhetoric to be the suasory use of language as a symbolic means of inducing cooperative acts and attitudes in symbolizing beings.[7] Rhetorical acts present allegations about what is; they symbolically advance contentions about how we should name, pattern, or define experiences and thereby make those experiences meaningful.

When considered as symbolic inducement, rhetoric explodes traditional boundaries. The entire range of human culture and experience now opens for rhetorical investigation. For example, the human sciences have formulated specialized vocabularies for their inquiries. When their investigations contribute to understanding of that dimension of symbolizing that involves inducing audiences, they contribute to studies in rhetoric, regardless of the terms they use. Rhetoric's range extends beyond public oratory to tribal rites and rituals, to the mystifications found in courtship and class relationships, to an individual's personal sense of place within a decaying, conflictive, or cohesive social order, to how individuals work toward maintaining, or transforming, particular social orders, and so on. The human sciences pursue rhetorical understandings of a wider range of symbolic phenomena than was envisioned by classical theorists, albeit often unwittingly and tacitly. Burke explains:

> Though rhetorical considerations may carry us far afield, leading us to violate the principle of autonomy separating the various disciplines, there is an intrinsically rhetorical motive, situated in the persuasive use of language. And this persuasive use of language is not derived from "bad science," or "magic." On the contrary, "magic" was a faulty derivation from it, "word magic" being an attempt to produce linguistic responses in kinds of beings not accessible to the linguistic motive. However, once you introduce this emendation, you can see beyond the accidents of language. You can recognize how much of value has been contributed to the New Rhetoric by these investigators though their observations are made in terms that never explicitly confront the rhetorical ingredient in their field of study. We can place in terms of rhetoric all those statements by anthropologists, ethnologists, individual and social psychologists, and the like, that lean upon the *persuasive* aspects of language, the function of language as *addressed,* as direct or roundabout appeal to real or ideal audiences, without or within.[8]

The idea that rhetoric is symbolic inducement arises from recognizing the fundamental capacity of human beings to make and respond to symbols. Humanity's nature can be defined in many ways, but the ability of human beings to argue about definitions of themselves points to their symbol-using (-misusing, -making) capacity as central among the differentiae that identify the genus "animal" when it refers to humans. Since humans are the kind of animal that is able to debate about the appropriate definition of its kind, the source of this ability must be ontologically prior. It is for this reason that Burke turned to humanity's symbolicity as that which designates its most universal classification, what Cassirer called *animal symbolicum.*[9]

At first glance the statement that a human being is a symbol-using (-misusing, -making) animal may appear a truism. Yet reflection reveals important implications for what we mean by "reality."[10] According to this view, reality is nothing other than that which is mediated through our symbol systems. We cannot think without symbols (unless we consider certain kinds of meditation as modes of thinking that can be used to experience reality nonsymbolically).[11] In any case, without the symbolic constructions of history, biography, law, art, poetry, music, philosophy, science, what would remain of our cultural traditions? Strip away the symbols that mediate what one remembers about the past, what we have learned through personal relationships, through books, magazines, newspapers, and through various electronic media, and what is left as a remainder of what we commonly refer to as the self? The central point is that because the human being is a symbol-using animal, it names its experiences, and through this symbolic act it creates, to a great extent, what it takes to be its world.[12]

Biologically, humanity can never fully escape the natural world that prescribes the boundaries of animal experience, but the capacity to symbolize enables humans to transcend purely materialistic nature. Humans enter a unique and complex experiential dimension through possessing the capacity to symbolize. Once we can *name* things and experiences, the universe takes on new qualities and new significances. As the cherubim with flaming swords forever kept Adam from Paradise, so does the mediation of symbols stifle humans' attempts to capture a state of nonsymbolic oneness with nature.[13] For good or ill, humanity cannot but accept the conditions of the symbolic world that it must live in.

Given the various other intellectual disciplines that contribute to studies of symbolic structures (e.g., logic, linguistics, poetics, hermeneutics), what is it about symbolic structures that makes rhetoric important in the study and practice of symbolic activity? In Burke's terms, what precisely is the "essential function of language itself"[14] that rhetoric accounts for? The answer is direct: Rhetoric explains the *selective* functions involved when we make, apply, and judge symbols. A symbolic actor can only exercise his or her symbolic capacity by selecting symbols through which to mediate experience and render it meaningful.[15] Furthermore, selection is necessarily persuasive in its consequences. In choosing one term over others, one directs attention toward particular meanings and relationships and excludes or minimizes those supplied by other terms. An individual selects words; they invite an audience (including the self) to conceive, analyze, and evaluate phenomena in a particular way. Should an audience accept this invitation to perceive phenomena through the "terministic screen" of a particular set of words, alternative symbols and their concomitant meanings will thereby be deflected from thoughtful consideration. This is one sense in which choosing among symbols is always persuasive (i.e., rhetorical). There is, however, a further sense in which all symbolic choosing is persuasive because selective.

When we choose what to say, our choices of words are inevitably made from our own perspectives on whatever ideas, things, and events we choose to discuss. Seldom if ever is there only one way to "see" a phenomenon. However, we symbolize phenomena in the ways we "see" them, and that is not necessarily the only way the phenomena can be "seen." All symbolic choices thus argue for some point of view concerning the phenomenon symbolized. Chaim Perelman indicates why this is inevitable: "If a word already exists, its definition can never be considered arbitrary, for the word is bound up in the language with previous classifications, with value judgments which give it, in advance, an affective, positive or negative coloration."[16] By selecting certain terms rather than others, then, we emphasize particular meanings and values—those that seem appropriate, given our understandings. This is part of the reason that Richard Weaver could aptly call rhetoric an "art of emphasis embodying an order of desire."[17] Linguistic choices request that an audience accept the terms *we* think are appropriate, together with their concomitant beliefs and values. There is nothing insidious about this. All languaging and symbolizing has to be selective, but given the myriad options symbolizing allows, every selection makes some problematic claim. Weaver, therefore, was on the mark when he said "language is 'sermonic.' " All users of language are preachers insofar as their choices of words bring into view the values, meanings, and purposes that they desire. By not choosing other verbal options, they exclude alternative values, meanings, purposes. The point is that every selection of a word through which we mediate a phenomenon is necessarily a suasive act that imputes particular meanings and deflects consideration of others.

Explicitly or implicitly, a terminology functions as an exhortation to think, act, and feel according to particular patterns prescribed and suggested by the terms.[18] Consider for example the academic who views with trepidation the meeting of a committee charged with determining whether he or she should receive tenure. Suppose this person were to say: "I know that at least three of the five committee members are in my foxhole." This statement can be shorthand for a comprehensive set of attitudes and values that constrain the academic's view of an ambiguous and potentially threatening professional situation. The statement quite probably reveals a view of the tenure process, not as a deliberate application of standards to the candidate's record, but as analogous to a struggle for military power or a venture in buddyism, or both. The implications of the candidate's choice of terms can predict further vocabulary. The candidate probably feels embattled, or caught in a struggle between warring factions. In any case, the terministic screen chosen in making the original statement betrays the candidate's general sense that strength and hope are secured through alliances with those who will "fight" or "protect" one's side in the "trenches."

The images of embattlement or of factional struggle that are suggested by the hypothetical statement almost surely reflect a vision of academic politics, but do they signify a vision that applies outside academic life? They could be reflective of a world view or general orientation toward "reality." When terministic screens reflect sweeping, suprasituational outlooks, they draw meanings from systematic relationships that *terministic orientations* supply. Terministic screens reflecting essentially transient, situationally determined viewpoints must be distinguished from those embedded in a terministic system of meaning relationships. The distinction is important because we often see cross-situational rhetoric that seeks terministically to gain adherence to complex, systematic, cross-situational understandings that are symbolized through a *system* of terms. In speech communication, for example, the term "reticence" symbolizes an approach to fear of speaking that in virtually all its terminology emphasizes the perceptual and cognitive conditions of the "shy," whereas the term "communication apprehension" identifies a significantly different approach to shyness—one that emphasizes the behavioral conditioning of the "shy." Conflict about such terms is more than quibbling about words. Each expression invokes its own distinct, underlying orientation that makes discussion of any specific instance of shyness fully meaningful.

When terministic screens of communication reflect systematic, cross-situational outlooks toward specific subjects, terministic orientations will constrain ongoing discussions.

Where a terministic orientation is coherent, three major sets of implied valuings exist:[19] instrumental valuings, logical valuings, and moral valuings. A coherent terministic orientation can reveal to an observer how a person using that orientation is likely to choose means for solving problems under a given set of circumstances.[20] The terminology implies principles of what is "appropriate" and what is "adequate" when a difficulty is addressed. We can see this clearly exemplified when we look at what is appropriate and adequate for solving problems humanistically or scientifically. Humanistic terminology leaves the way open for alternative interpretations; scientific terminology expresses precision and certainty as regulative ideals. In a paper titled "The Language and Methods of Humanism," M. H. Abrams contrasted the criteria of appropriateness and adequacy reflected in scientific and humanistic orientations when he said:

> To demand certainty in the humanities is in fact to ask for a set of codified rules and criteria such that when, say a work of literature is presented to any expert critical intelligence, it will process the work and come out with a precise meaning and a fixed grade of value that will coincide with the meaning and evaluation arrived at by any other critical intelligence. When a humanist really faces up to these consequences of his demand for certainty, he finds such a mechanical process to be disquieting and repulsive—and with good reason—because in fact it is approximated [in humanistic matters] only under an authoritarian cultural regime in which the codified rules and universal criteria are not discovered . . . but are established by edict. In our free humanistic activities, we all take for granted the human predicament that the humanities both deal with and express, and we manage as a matter of course . . . to cope with a situation in which tenable perspectives are diverse, individual sensibilities and proclivities are distinctive, many judgments are contestable, and few basic disagreements are in any final way resolvable. . . .
>
> As a matter of everyday practise, however, we in fact possess various ways for checking our individual judgments and for establishing the difference between the better and the worse, the greater and the lesser, by criteria that transcend our personal predilections and judgments. Chief among these is a revised form of what used to be called the *consensus gentium*. . . . We can state the principle in this way: agreement

> among diverse humanists as to the importance and value of a work at any one time, and still more, the survival value of a work—general agreement as to its importance and value over an extended period of time—serves as a sound way to distinguish the better from the worse and to identify which work is a classic.[21]

Differences between scientific and humanistic approaches to the study of human affairs are, in large part, differences of views concerning the most appropriate interpretations with which to discover what problems are important and concerning what instrumental methods are adequate for addressing important problems with efficacy. One need not know a communicator's history to observe whether the terministic screens he or she invokes define problems from a humanistic or a scientific orientation.

A terministic orientation also imposes an internal logic on choices and on the structure and development of their presentation in consecutive discourse. As Burke explains, "We are logical (logos: word) when we specifically state the nature of a problem and then go and see within the terms of this specific statement."[22] It can be added that we not only "see" within the terms of specific statements concerning what is at issue, we *reason* in accordance with the criteria for legitimacy that are implied by that way of seeing. Thus rhetoric acquires its informal or, as I shall later call it, "topical" logic. For example, when the father of behaviorism, John B. Watson, contended that explanation of human behavior could be derived by objectively observing what the human organism says and does, his statements prescribed boundaries for meaningful talk consistent with his orientation toward psychology. As he put it, the problem for the behaviorist was this: "Can I describe this bit of behavior I see in terms of 'stimulus and response'?"[23] Once the problem was stated in this way, certain terms and ways of thinking became "mediaeval": "sensation," "image," "purpose," "thinking," "emotion." These concepts and the lines of reasoning they allow and suggest must be swept aside as useless ways by which to comprehend human behavior. All of this was logical, given the initial terms of Watson's orientation. However, what Watson and others could not do was charge those who saw meaning as residing in intentions or purposes with being irrational or illogical. Given their terminological definitions of the problem, they too were logical in perceiving and reporting behavioral phenomena. What Burke called terministic orientations vary in "doing" psychology and each orientation can prescribe different premises for thinking about and discussing psychological matters. Watson's orientation decreed that only observable behaviors counted as scientifically reasonable data. William James's orientation allowed additional kinds of data to count as scientific—for example, self-reports of feelings and attitudes. The logic of rhetoric in psychology rests not only on formal criteria for inference but on informal criteria of "legitimacy" applied to data and inferences. Psychologists adhering to different orientations may make incompatible or conflicting claims, but nevertheless each can be "moving from premises to conclusion with the syllogistic regularity of a schoolman."[24] Thus, at least two kinds of informal logic operate in rhetorical discourse: the logic of terminological choice and the logic that prescribes what counts as legitimate data and inference.

Third, any terministic orientation is a framework for moral evaluation. Placing a problem within the context of an orientation invokes a vocabulary of ought and ought-not, of what is praiseworthy and blameworthy.[25] For instance, a scientist acquires through professional education and training notions of what it means to think and act as a scientist. If a scientist is convinced that knowledge claims should be certified against preestablished, impersonal criteria that can show those claims are consonant with observations and with previously confirmed knowledge,[26] that scientist is likely to consider "objectivity" or "emotional neutrality" a necessary characteristic of what it means to be scientific. The scientist will strive to act accordingly. As terministic orientations toward what it means to be scientific shift, there will be shifts in the moral frameworks for interpreting appropriate and inappropriate scientific behavior.[27]

In sum, a terministic orientation provides makers and observers of discourse with a complex of interwoven meanings that prescribe and constrain instrumental, logical, and moral choices. Some caution is needed at this point, however. We should not start to conceive of terministic orientations as free-floating structures wholly separate from experience. The strength of an orientation and the interpretations it provides are influenced by how well the orientation accounts for (and mediates) the contingencies of experience. A scientist may be scientific about nature but humanistic about public duties. The terms of an orientation supply heuristic classifications that enable a user to interpret experience in a particular way. A terministic orientation gives an order to some body of experience, but other options are always open.

The strength of an orientation rests on the capacity of its classifications to provide a sense of what one should expect in a situation. Should orientation-induced expectancies not be fulfilled in one's experience, the user is likely to become disoriented, unable to come to terms with the situation.[28] The serviceability of the adopted orientation would then come into serious question. As need for a reorientation or shift of perspective becomes manifest, the consequence is likely to be a war of words. Rhetoric will be used to allege, through different sets of terms and different "rules" of evidence and inference, how events ought to be classified to restore order. It does not matter whether a shift in perspective is disputed in politics, religion, or science; the dispute is equally an exercise in rhetoric.

Rhetoric is also used to mend differences and clarify confusions among those trying to extend a shared orientation and, especially, among those working from different orientations. Efforts to bridge gulfs of division among orientations, or to expose those gulfs as illusory, are always rhetorical. It is obvious, then, that rhetoric's arenas are far more inclusive than those of public oratory of politics and law.

Those who make interpretations of phenomena from different orientations are not likely to achieve lasting identification with each other. In differently naming and thereby characterizing an object of discussion, each determines its *substance* differently and therefore disagrees with the other as to its meaning. These disagreements might also expose deeper conflicts over the principles each employs to legitimate claims when arguing with others. Conversely, when a speaker and audience share a terminology and underlying criteria for logical valuation, they may be said at least to begin to agree on meaning. A rhetor must therefore strive to find terms that indicate sharable sensations, concepts, images, ideas, and attitudes.[29] This is operational pursuit of what Burke calls consubstantiality.[30]

We make rhetoric to induce others to cooperate with us in thought and deed, to participate in our ways of looking at a situation and its relationships, to share our terministic orientations. In making such attempts we consider what resources are available for transcending division and achieving identification. We may or may not assess available resources fully or wisely, but whatever resources are available varies across situations.

RHETORIC IS SITUATIONAL DISCOURSE

Aristotle defined rhetoric as the faculty of discovering the available means of persuasion *in the given case*.[31] Ever since, most rhetoricians have presumed that some relationship exists between rhetorical communication and the types of occasions from which or in which it emerges. A classical orator was taught to extract topics for speech by locating materials that "fit" the deliberative, forensic, and epideictic arenas of oratory.[32] Today, of course, rhetorical occasions exceed in number and complexity those that were systematically explored by classical rhetoricians. Rhetoric in the modern world requires a view of rhetorical situation that extends beyond public oratory to contexts wherein scientists deliberate, academics argue, and to a number of other types of forums.

Not all situations are rhetorical, of course. How are we to recognize a *rhetorical* situation when we encounter one? Lloyd F. Bitzer has identified a special mix of features that together mark a situation as rhetorical or one in which rhetoric is potentially relevant:

> Rhetorical situation may be defined as a complex of persons, events, objects, and relations presenting an actual or potential exigence which can be completely or partially removed if discourse, introduced into the situation, can so constrain human decision or action as to bring about the significant modification of the exigence. Prior to the creation and presentation of discourse, there are three constituents of any rhetorical situation: The first is the *exigence;* the second and third are elements of the complex, namely the *audience* to be constrained in decision and action, and the *constraints* which influence the rhetor and can be brought to bear upon the audience.[33]

The controlling feature of a situation that invites rhetoric is the exigence. An exigence is a problem that emerges in an interpersonal context. Something is perceived as other than it should or could be. The problem can be an ambiguity, defect, or obstacle. The imperfection is a *rhetorical* exigence when it has a contingent nature, and so is subject to modification or removal through alterations of an audience's acts and/or attitudes. For rhetorical discourse to address an exigence effectively, it must constrain and guide an audience's decisions and actions. A *rhetorical* audience is one that is capable of being influenced by discourse and capable of actually mediating change in respect to the exigence. Such an audience must be present for a situation to be rhetorical. Finally, rhetorical situations include preconditions that constrain speakers or writers as they attempt to respond fittingly to the exigence. These constraints include preexisting behaviors, attitudes, documents, facts, traditions, images, interests, and motives of participants. *Rhetorical* constraints limit or enhance opportunities for making appropriate rhetorical responses to an exigence.

Especially pertinent to concerns of this book is the fact that for scientists and academics, a rhetorical exigence can be a gap in the collective body of knowledge. Perhaps a pocket of data within a particular domain is not accounted for adequately by current theory. Informed scientists or other scholars become aware of this and it becomes an exigence. They see it as a need to locate propositions that enable explanation, prediction, and control of the phenomena at issue. Such an exigence becomes a rhetorical exigence when a scholar attempts to convince interested colleagues that he or she has a solution. Like any other rhetor, the scholar proposing a solution needs to adjust appeals to the constraints of the rhetorical situation. If wise, he or she will consider the available terministic orientations that exist in the situation. Those orientations prescribe how to state the problem, how to proceed toward its analysis and resolution, and what will be the criteria for evaluating proposed solutions. The rhetorical audience—members of the community of interested scientists or other scholars—will judge the results of the proponent's reflections and decide whether the exigence has been resolved or significantly modified. Judgments about the fittingness of a scholar's response to an exigence are likely to take place at professional meetings, during editorial discussions of whether to publish or not to publish results, or in private or public discussions about results once they are published.

Through rhetoric one can also create rhetorical situations, or influence interpretations of their features, as a prelude to addressing the exigences thus brought to attention and terministically defined. To induce others to cooperate by addressing exigences from within the framework of a particular terministic orientation may itself be an intermediate exigence that requires rhetorical address. Such appeals for cooperation are often of crucial importance because any given terministic orientation presupposes or prescribes special meanings derived from naming the features of the ultimate rhetorical situation, defining how these features relate to and influence each other, and invoking criteria for determining success or failure in resolving the problematic situation. M. H. Abrams's discussion of scientific and humanistic standards for inquiry illustrates this point. A scientist's terministic orientation places degrees of certainty at issue; a humanist's terministic orientation places qualities such as originality, arguability, and consensus at issue. A scientific orientation inclines one to define rhetorical situations in terms of the presence or absence of objectively verifiable information and the consonance of new evidence with already-accepted knowledge. A humanistic orientation inclines one to define rhetorical situations in terms of possible interpretations. Abrams's remarks themselves involve inducing scholars to adopt a humanistic rather than scientific orientation toward problems in literary, historical, and philosophical studies. Generally, attempts to solve problems may call for rhetoric that tries to induce others to share one's terms and implicit values before one tries to address the ultimate exigences of a situation.

All rhetors are further constrained by historical circumstances.[34] Cooperative acts or attitudes are always induced in situations that have a past, a present, and a future. It is obvious, too, that historical constraints transform over time: The worlds of Copernicus, Galileo, and Newton were not the same as the world of contemporary science, and scientists are aware of this. Consonance of discourse with constraining historical circumstances will not guarantee rhetorical success, but appropriateness to past and present is a necessary condition for effective inducement.

To maximize the potential for persuasive success in a rhetorical situation, the rhetor must engage in rhetorical actions that are, and will be perceived by an audience as, purposive. Unintentional, accidental, semiconscious, or subconscious modes of activity may be instances of symbolic inducement; however, to maximize persuasive opportunities, rhetoric needs to be purposively designed and aimed at specific goals. To succeed, rhetorical discourse must reconcile the personal and private inclinations of a rhetor with principles acceptable to the audience. The apparent purpose of such discourse must suit audience expectations in the situation. Accordingly, a rhetor must pursue purposive reflections directed toward analyzing the persons and groups to whom discourse is to be addressed.

RHETORIC IS ADDRESSED DISCOURSE

Donald C. Bryant has described the function of rhetoric as "the function of adjusting ideas to people and people to ideas."[35] Any concept of rhetoric must recognize this adaptive function—its nature *as addressed.* Symbolic inducement occurs through both public and private discursive address. In antiquity rhetoric was usually correlated with public address in practical forums. Some contemporary theories extend this idea to intrapersonal ruminations. When a person addresses his or her own self by cultivating, in secret thoughts, images and ideas, that person is being as rhetorical in self-inducement as an orator who uses imagery to persuade public audiences to adopt a particular point of view.[36] Our self-inducement yields "audience" responses that help us (1) maintain our "selves"; (2) form judgments about practical problems that confront us during day-to-day living; (3) invoke acceptance of attitudes, ideas, and feelings that help us to render actions purposeful; (4) rationalize our errors; (5) assuage or exacerbate guilt feelings; and (6) bolster or diminish ego.

Symbolic inducement also occurs, of course, when rhetors address public audiences identified by traditional rhetorical theory. The political parliament and populace, the judge and jury, the religious congregation, are all capable of symbolic inducement. Classical rhetoricians, however, treated public rhetorical audiences as simply given.

In the literature of antiquity it was not uncommon to distinguish between dialectic, or expert discourse addressed to a few, from rhetoric, or popular discourse addressed to many. This distinction is more typically rejected in contemporary theorizing.[37] Persuasion of any audience is rhetorical action, and wherever judgments are involved, there will be persuasive efforts. Those who judge the worth of expert discourse are no less rhetorical audiences than are legal or political audiences; rhetors in either context make discourse that they hope will constrain a judging audience's decisions and thereby induce favorable verdicts. In all such cases rhetorical discourse is used to address exigences, and the success of the discourse depends on persuading situated, judging audiences who are sensitive to those exigences.

Claims to knowledge require communal authorization before they become officially incorporated into a community's stock of knowledge and problems for professional inquiry. Both those seeking accreditation for claims and those empowered to authorize or test those claims purportedly engage in these rhetorical activities in behalf of the knowledge community to which they all belong.[38] Each community has its own orientational, "logical" system for admitting and evaluating claims. Accordingly, an authorizing audience adjudicates claims put forward for accreditation with reference to interests, terms, values, concerns, concepts, and techniques allegedly shared by most, if not all, members of the knowledge community they represent. Those seeking accreditation rhetorically adjust and adapt their claims to the communally sanctioned standards and corollary expectations both rhetors and adjudicating audiences allegedly share. Claims to knowledge thus gain accreditation from an expert audience by such means as rhetorical display of data in accordance with received tests of relevancy, technical skill in developing and expressing arguments that are warranted by shared community values, and application of claims to the problem-solving concerns and troublesome issues that confront the knowledge community. Decisions to authorize or not authorize such claims are based ultimately on communal standards for adjudicating the reasonableness of any given rhetorical performance.

In contrast to the concepts "valid" or "correct," the concept "reasonable" allows us to recognize that a favorable judgment about a claim may, but need not, involve commitment. There is a whole range of attitudinal dispositions that an audience can have toward claims, all short of commitment but still allowing formal endorsement of the discourse's reasonableness in a given situation. For example, consider an academic rhetor who seeks publication in a respected journal or acceptance of ideas for presentation at a professional conference. It is likely that this rhetor hopes to persuade an audience of reviewers that his or her proferred ideas are acceptable because they engage technical exigences appropriately and usefully, given the standards and concerns of the community that reviewers allegedly represent. Optimally, the rhetor would want to gain commitment to the ideas, but it is not likely that he or she will achieve faithful and total commitment. It is more likely that an appreciative understanding of the case presented will constitute the grounds for accepting a paper submitted for review. Judgment of a paper's acceptability is not as much a matter of its "truth" or "correctness" as an acknowledgment that the rhetor has developed arguments that are plausible enough to deserve a hearing by a professional audience. Reviewers' personal opinions and commitments may or may not be identical with the ideas in a presentation they approve for publication. Accordingly, we may say the reasonableness of discourses submitted for consideration becomes a judgment expressing the degree to which the discourses evoke appreciative understanding as contributions to ongoing discussion of professional problems.

Of course, audiences of specialists such as those referred to do not comprise the only audiences who respond to scientific or other professional discourse. There may be a number of interested audiences for any particular discourse—audiences having different levels of competence to judge understandingly. For example, in scientific activity the fully qualified audience may, on some technical matters, consist of not more than a dozen trained scientists working to resolve exigencies that the discourse addresses. This small, specialized audience would have the decisive role in authorizing the claims to knowledge. On the other hand, if the discourse successfully defends some special achievement with broad social implications, the information would probably be recast in more common parlance for audiences with lower levels of technical competence.

If it were published in a general science magazine, intelligent laypersons could read about the achievement and judge it on their terms. They would not be the active authorizers of the knowledge claims, but they would nevertheless judge the work as interested witnesses. Laypersons might mediate changes in culturally received information, but they would not give technical knowledge claims their technical standings. A technical audience would do that.

It is not always obvious which audience is the authorizing audience and which is a less technically competent but still culturally important audience. There may even be conflict and ambiguity concerning kinds of relevant competency. For instance, there has been much public discussion of the Strategic Defense Initiative, more popularly known as the "Star Wars" laser defense system. Whose standards of reasonableness should be used to supplement political standards when active members of the public seek to warrant acceptance or rejection of continued research on this system? Scientists'? Economists'? The military's? The clergy's? Whoever makes sound rhetorical judgments of such discourse must discriminate among available standards of rhetorical reasonableness, evaluate their comparative importance given the issues, and determine which ones singularly or in some combination operate (or should operate) to constrain the boundaries of legitimate debate.

The source of authorization for any claim is found in standards of reasonableness that reflect the general interests and common concerns of a community. It is to be expected, then, that theologians, scientists, lawyers, would stand for different communities of interest and thus authorize claims in accordance with varying standards of reasonableness. Whether a rhetor makes scientific, theological, or legal claims, utterances must defer to the professed standards of reasonableness held by the particular audience addressed and charged with authorizing powers. There is no successful rhetorical communication otherwise. Given this or that public issue, professions of theological truth, references to scientific facts, and appeals to legal precedent will likely seem reasonable to some audiences and unreasonable to others. Understanding of these differences and the reasons for them is a contribution that rhetorical analysis of discourse can make. In any case, a rhetorical analysis does not (and cannot) determine the truth of a claim; the rhetorical issue is how and why the discourse succeeded or failed to meet the standards of reasonableness held by the audience(s) that responded to it.

To succeed rhetorically, any rhetor must ground a position in what the addressed audience considers reasonable; only then can understanding and appreciation be achieved, and only then can approval and authorization even be hoped for.

RHETORIC IS REASONABLE DISCOURSE

The idea of rhetorical reasonableness dissolves dichotomies so often imposed by commentators on science. The tendency to divide discourse into "rhetorical" elements that are emotional, stylistic, or even irrational and "nonrhetorical" elements that are intellectual, substantive, and rational is based on nonrhetorical standards for determining the plausibility of ideas. According to Wallace, rhetoricians have been trapped into similar dichotomizing by their uncritical acceptance of a scientific realism (or naïve realism).[39]

Wallace urged that rhetoric be conceptualized as the art of finding and effectively presenting "good reasons." For Wallace, good reasons, which are drawn from moral values, constitute the "substance of rhetoric." A friendly extension of Wallace's view would be to argue that rhetorical discourse is discourse *made* reasonable.[40] This view avoids unproductive dichotomies like those outlined above.

Perelman offers a distinction between the rational and the reasonable which is useful as a starting point in delineating that which is reasonable. The rational is valid only in an abstract theoretical domain and is characterized by rigid adherence to a priori standards.[41] The reasonable, in contrast, is concerned with making judgments and engaging in actions that are commensurate with the dictates of common sense. Perelman elaborates the tendencies of the reasonable man as follows:

> He is guided by the search, in all domains, for what is acceptable in his milieu and even beyond it, for what should be accepted by all. Putting himself in the place of others he does not consider himself an exception but seeks to conform to principles of action which are acceptable to everyone. He considers as unreasonable a rule of action which cannot be universalized.[42]

Perelman's distinction is useful if one keeps in mind what Perelman calls "particular audiences." There is an unsettling tendency in Perelman's writings to stress reasonableness as what is judged reasonable by a "universal audience" that transcends any particular grouping. This has been the most seriously challenged aspect of his theory of argumentation,[43] but when he grants that the reasonable varies from age to age and that it is tied closely to the commonly held opinions of particular audiences, his distinction between the reasonable and the rational becomes fully consonant with the concept of rhetoric as *all* suasion intended to alter attitudes and actions.[44]

If we recall Burke's concept of terministic orientation, we are in a position to see how and why a rational perspective may be rhetorically *unreasonable*. Insofar as an individual works consistently within an orientation, his or her theoretical reflections or practical behaviors may be said to be rational. The reflections and behaviors will follow the logic prescribed by that orientation. If that person uses symbols that induce others to participate within the adopted terministic orientation, the resulting rhetorical discourse will be reasonable as well. That is, the speaker or writer will in some way intersect his or her perspective with the common opinions and logical principles of the audience and will in some way address demands of the situation. On the other hand, to neglect the concerns and interests of the audience and the demands of the situation while attempting to induce altered attitudes or actions would result in unreasonable discourse, regardless of the rationality of the perspective out of which it developed.

Discourse must be reasonable to be rhetorically significant. Fisher provides a clear definition of that which is accepted as reasonable by an audience in a given situation.[45] For Fisher, "good reasons" are *"those elements that provide warrants for accepting or adhering to the advice fostered by any form of communication that can be considered rhetorical."* "Warrant" in this usage is "that which authorizes, sanctions, or justifies belief, attitude, or action—those being the usual forms of rhetorical advice."[46] The notion of "good reasons" does not, however, mean that any element that warrants some belief, attitude, or action is as good as any other warranting reason. "Good reason," in Fisher's view, "signifies that whatever is taken as a basis for adopting a rhetorical message is inextricably bound to a value—to a conception of the good."[47] Fisher's conception of good reasons is useful in making clear that when rhetoric is made about controversial and contingent matters, the discourse must be grounded in positive standards of value if it is to be judged reasonable. When audiences make judgments about the reasonableness of particular and disputed claims, they will decide on the basis of whether the rhetors furnish good reasons for those claims; they will look for grounds to help them determine which claims warrant assent or dissent. When rhetors choose and develop good reasons for their claims, they encourage their audiences to judge controversial or ambiguous beliefs, attitudes, or actions by using those comparatively more stable and less controversial good reasons as evaluative criteria. In this way rhetors try to give their audiences standards for deciding that particular claims are situationally reasonable.

Where and how can a rhetor locate rhetorically efficacious reasons for warranting acceptance of claims as reasonable? The single most important source is the audience. Arnold stresses the centrality of the audience in determining what is and what is not reasonable when he offers this definition of the reasonable:

> In general communication, "reasonable" and "logical" occur as terms *people* use when they want to report that something "hangs together" for *them,* seems adequately developed for *their* purposes, seems free of inconsistencies insofar as *they* noticed. It seems to me we *must* admit that in rhetoric and drama at least, "reasonable" is in the final analysis what the consumer is willing to call "reasonable."[48]

Arnold then elaborates by observing that a general characteristic of men and women is their tendency to hunt for interconnections among ideas and things, especially when they are uncertain or begin to doubt. I propose that those opinions and judgments that help an audience to establish interconnections and resolve matters in doubt are the good reasons of rhetoric. However, I want to avoid Arnold's tendency to slip into the rational-irrational (logical-illogical) dichotomy, as he does in opposing the challenges of the "New Romantics" in respect to reasonable talk.[49] By recognizing with Arnold the importance of the opinions and judgments of an audience as the source of that which is reasonable, the likelihood of raising such unproductive dichotomies is markedly reduced. A reasonable rhetor need only operate on this principle: If in the opinions of a given audience certain kinds of conduct or belief are reasonable, then to be reasonable the rhetor must identify his or her position with at least some of those kinds of conduct or belief.

Audiences' opinions of what is reasonable will, of course, vary with the historical and social contexts within which they live. I have already observed that the historical circumstances in which a rhetorical situation occurs constrains rhetors' selections of symbols with which to alter that situation. This fact has both macrocosmic and microcosmic applications. Historical circumstances affecting all but a very few rhetorical situations in the twentieth century are unfriendly to spokespersons for the Flat-Earth Society. The dominant ways of thinking consign their claims and arguments to the realm of the unreasonable. The opposite is true of almost any imaginable situation in which a rhetor backs up claims with evidence from empirical experimentation. We are used to that as a reasonable kind of "proving." Putting the point generally, the zeitgeists of historical epochs set boundaries for reasonableness on some subjects at some times. More particularly, discourse that is considered reasonable has to be consonant with the specific opinions, judgments, and concerns that make any situation a rhetorical situation. The practical possibilities available for evoking an exigence, a constraint, or a rhetorical audience are thus defined by both the propaedeutics of the era and the immediate and specific historical circumstances of every particular audience.

It is now possible to state with some precision the standards by which rhetorical audiences judge or authorize discourse. The standards are communally shared standards of what is "reasonable." The standards are fluid, being under the influence of historical, situational forces. However, because they are shared standards, they also have considerable stability, solidity, and endurance. They are not personal, idiosyncratic rules. They reflect commitments and expectations of a community, however enduring or temporary that community may be. Insofar as various audiences have similar concerns, exigences, constraints, and cultural presuppositions, they share standards of reasonableness. However, particular communities of interest and expectations also exist, resulting in special standards of what is reasonable, often more stringent than the standards of a general culture. This, in sum, is the general nature of the grounds and principles that govern the informal logic of all rhetoric.

The reasonable, so understood, fuses at one stroke the dichotomies of structural and substantive features of discourse. A reasonable discourse must, of course, display patterns of organization and style familiar to the situated audience. The discourse must connect with the shared beliefs, values, and attitudes of the audience if they are to find its reasons "good." Appreciative understanding or acceptance depends on the rhetor's ability to locate, in form *and* content, ideas and images with which the situated audience can identify and thereby find reasonable.

RHETORIC IS INVENTED DISCOURSE

Traditional rhetorical theory recommends three inventional procedures for thinking rhetorically about discursive content. First, a rhetorical goal or purpose must be isolated. Discourse will have coherence and focus only if it is organized around a clear and specific aim. This is an important first step because determining the appropriateness of content depends in part on the purpose or purposes sought through rhetorical communication.

As I shall show in the next chapter, inventional principles provide method in locating these ends.

Second, a rhetor must be able to discern what is at issue within a rhetorical situation and how content pertinent to the issue or issues should be ordered. It is therefore necessary to define the questions at issue and to discover content most suitable for rhetorical intervention at the relevant points. Classical rhetorical theory addressed these two problems by developing what was called a *stasis* procedure. Social experience has become more varied and complex in modern times, but it remains true that inventional procedures are needed to help determine which issues do and do not deserve treatment. I shall discuss those procedures in chapter 4.

Third, any rhetor has to decide what to say about points at issue. Inducement of acts and attitudes requires content that is relevant and appropriate to the most significant aspects of the now-narrowed subject. Classical rhetorical theorists suggested *topoi,* or sets of topics, as heuristic aids for finding appropriate sayables about specific points at issue. *Topoi* still supply this guidance today, helping rhetors canvass and choose among possible sayables about a subject or problem. Chapter 5 treats this subject.

A general theory of rhetoric that takes advantage of traditional and contemporary thinking about the three sets of problems just identified shows how one can think rhetorically about the content of discourse whatever the subject. It is possible to extract from general theory special theories of rhetorical invention appropriate to specific communities of thought and opinion: science, sociology, history, philosophy, or any other substantive intellectual discipline. To do so, we must (1) explain the general theory of rhetorical invention, and then (2) extract and apply a specific theory of science as rhetoric.

3

RHETORICAL PURPOSES

It is a commonplace that rhetorical communication must be oriented toward "purposes,"[1] but theorists have explained the principles for determining and pursuing the purposes of rhetorical discourse in two distinct ways. One line of thought has been developed on the basis of psychological presuppositions about the general nature of human minds and responses. An alternative approach has been developed from presuppositions about special kinds of situations conducive to rhetoric. In this chapter I shall examine these two approaches in order to extract principles about purposing that can most usefully guide processes of inventing rhetorical communication.

PSYCHOLOGICAL APPROACHES TO RHETORICAL PURPOSE

Contemporary rhetorical studies have been illuminated and informed by different psychological schools of thought. Rhetorical critics have used psychological categories to probe and reveal deeper textures of meaning, exposing psychic motives, impulses, and processes that lurk beneath the surface manifestations of rhetorical practice.[2] Rhetorical theorists have sought firmer foundations for theoretical reflections by deriving and applying principles about mental processes from specific fields or theories of psychology.[3] Though generating useful critical and theoretical insights, psychological approaches have been strangely mute when asked for principles to guide inventional decisions involved in preparing and practicing rhetorical communication. Specifically, they have uniformly failed to provide guidance for thinking about how to choose, formulate, and pursue rhetorical purposes. In "psychologizing," rhetoricians presuppose, or at least have concentrated on, universal principles or laws that presumably govern the mind or brain. This leads to minimal consideration of specific, inventional adaptations that must be made if rhetoric is to succeed with particular people in particular situations.

An excellent and thoroughgoing example of psychological investigation of rhetorical purposing and of its shortcomings is the theoretical work of eighteenth-century rhetorician George Campbell. As he said in the preface to his *Philosophy of Rhetoric,* he began with the premises of what he took to be the universal principles of human nature.[4] His inquiry into the principles of rhetoric was premised on psychological doctrines which asserted that man has a natural, innate disposition to believe that which results from sense experiences. Naturally, then, he developed a theory of how discourse strongly or weakly engages human minds in rendering beliefs and feelings comparable in intensity with those originally rooted in sensory experience. When dealing with the substance of discourse, he emphasized the kinds of materials that engage the basic laws constraining the beliefs of *all* men. Rhetorical purposes and the discursive means for securing them were articulated according to discrete psychological faculties and powers of association.[5] Rhetoric was "intended to enlighten the understanding, to please the imagination, to move the passions, or to influence the will."[6] This emphasis on universal principles is nowhere more evident in the *Philosophy* than in his advice about how a message should be adapted to the minds of particular, specially situated hearers. He devoted only one page out of 415 to how discourse is to be adjusted to "men in particular!"[7]

Use of the broadest psychological principles to explain rhetorical purposes is no mere peculiarity of eighteenth-century thinking. Contemporary authors taking a psychological approach to rhetorical purposing disagree over details, but the standard treatments in many modern textbooks on persuasion and public speaking emphasize the general, psychological-attitudinal-behavioral responses a speaker intends to achieve through discourse. Twentieth-century theorists tend to draw upon the principles of behaviorism and to urge practitioners to discover discursive purposes by considering psychological principles that apply to all audiences. They minimize the situational features that differentiate particular types or classes of audiences. It is said that instructive speeches aim to inform hearers by achieving clear understanding; recreative speeches attempt to entertain by providing interest and enjoyment; and persuasive speeches strive either to stimulate audiences through emotional arousal, to convince by compelling intellectual assent or belief, or to actuate by motivational and intellectual appeals crafted to get hearers to *do* something.[8]

Once a general purpose is formulated, speakers or writers are typically advised to review sets of strategies and devices for producing this or that major kind of response. As with George Campbell, there is usually little advice about how to discover what is appropriate to a *particular* audience. Speakers and writers are told that convincing an audience requires concrete facts, vivid illustrations, and logical reasoning; stimulating hearers or readers demands vivid imagery and compelling motivational appeals; and inducing listeners to act requires a balance of appeals to the intellect and to the emotions.[9] There the counsel on rhetorical invention usually stops, or attention is turned to methods of library and other research.

This is the overall conception of rhetorical purposing one derives from psychological approaches to rhetorical communication. Practitioners' attention is directed toward an explicit set of universal tendencies. It is implied that a speaker or writer should choose as the controlling end of all that is said or done the *one* of these tendencies that he or she hopes predominantly to evoke through discourse. The possibilities of mingled response patterns or of shifting goals in various parts of a communication are, if not explicitly pronounced inappropriate, treated as exceptional cases or not treated at all.

This kind of analysis of rhetorical purposing directs attention away from specific, situational, and individual adaptations that must be made if rhetoric is to succeed with particular people in particular situations at particular times. In fact, it is only through considering particular situations and how those situations generate special problems and values that appropriate rhetorical purposes and the content most relevant for securing them can be located. The inventional tasks in making rhetoric are not to choose *a* type of general and abstract response and then to build discourse according to an implicitly prescribed plan; the tasks involve deciding what kinds of specific, substantive aims one wants to achieve, at what points in communication, and then to choose content that will, at appropriate stages of discourse, motivate responses in particular, situated people.

SITUATIONAL APPROACHES TO RHETORICAL PURPOSE

Rhetoric was viewed by Aristotle as a socially useful discipline for furthering community aims by influencing the thought and action of Athenian citizens. All rhetorical creativity was directed toward marshaling the means of persuading an audience that a particular "cause" furthered an accepted community aim: "Rhetoric may be defined as the faculty of observing in any given case the available means of persuasion."[10] We, too, should think of rhetoric as the central means of intervening in the lives of communities.

Rhetorical purposes are not determined on the basis of psychological principles that are true for all audiences; rhetorical purposing is grounded in patterns of judgment that particular kinds of audiences usually apply within particular kinds of circumstances. Thus, Aristotle distinguished kinds of public speaking according to the kinds of audience decisions normally sought and rendered within special, recurring types of situations.[11] In Bitzer's language, rhetorical purposing must make adjustments to constraints regarding what can be said fittingly in response to special kinds of recurring exigences, or ambiguities, or points-for-decision. We are dealing now in sociological-anthropological as well as in psychological principles.

In democratic Athens rhetorical audiences were charged with making decisions that clarified ambiguities or solved exigences concerning past, present, and future conduct. The kinds of decisions they made would suggest which rhetorical means ought to be selected as most likely to induce favorable decision making in the setting. If the audience had to decide whether a person was guilty or innocent of past wrongdoing, the forensic orator had to craft a speech that accused or defended that person. If the audience had to judge between alternative courses of future conduct as useful or harmful, political orators needed to frame speeches that exhorted for or dissuaded from particular courses of action. If the audience was going to be interested in the value of words and deeds that pertained to the present, a ceremonial orator sought to praise or blame particular individuals or communities.[12]

Within this way of thinking, rhetorical means are judged by whether an audience will be convinced that the position advanced furthers some cherished principle or value. A rhetor will search out a principle or value communally accepted as appropriate to the given situation and use that principle as the "rhetorical end" of his or her rhetorical effort. Aristotle identified three typical sets of rhetorical ends that an orator was likely to pursue in the public arenas of Athens: Political oratory sought to establish proposed actions as expedient or inexpedient, forensic rhetoric aimed at proving actions just or unjust, and epideictic discourse worked toward demonstrating an action as honorable or dishonorable.[13]

Any persuasive influence a rhetor has with any type of audience depends on that audience's acceptance of a principle, value, or "reward" as a good reason for acting as it is urged to act. Because of what the audience will be valuing, a legal pleader is likely to be ineffective should he or she argue for acquittal of a client on grounds of expediency. The kinds of audiences that see themselves as judging guilt or innocence will not be interested in expediency as a prime value in their situation. On the other hand, if the pleader can show that the principal end of justice can be served by acquittal and that, in addition, acquittal would also be expedient in this case, the judging audience is likely to find acquittal an appealing decision. Subordinating one rhetorical objective to another in this way is not a matter of form but of locating a persuasively relevant subtheme appropriate to, but not central to, the situation. The broad principle here is: A rhetorical goal or end is compelling only if it draws on standard or conventional patterns of thinking that will be employed by one's specific audience while that audience is evaluating discourse about a particular kind of situational exigence.

Any principle upon which purposing is based needs to exert a normative influence on speaker, audience, and their interaction within a particular kind of situation. Should discourse or action contravene conventional, or normative, patterns, members of the concerned audience are likely to respond with rejection or even censure. In contrast, discourse or action judged as affirming and furthering a normative principle will be approved or even rewarded. Thus a fundamental and minimal requirement of successful rhetoric is that an audience accept the rhetoric as furthering aims consonant with their normative principles of action in the kind of rhetorical situation they are experiencing.[14] Should a rhetor base a rhetorical purpose on a principle that is inappropriate for the audience on the given occasion, the rhetor's character can become subject to question. For example, if a statistician should justify having used a statistical shortcut on grounds that it is *easier* than a more powerful and thoroughgoing analysis, there is a sense in which he or she would simply cease to be a statistician-persuader for people who understand statistical methods. The example can be generalized: The typical normative principles accepted by audiences in specific kinds of rhetorical situations prescribe the minimum basis of friendly rhetorical response.

As soon as one knows what response one wants from an audience and on what problem, Aristotelian theory posits that one should look for the special *topoi* that identify conventional ways of thinking that the "judges" will use in making the sort of decision the rhetor is asking for. From Aristotle's point of view, the statistician should have recognized that thoroughness was a primary value or special topic scientists and others would use in evaluating conclusions derived from statistical analysis. The statistician should have seen that since his or her purpose was to have a statistical conclusion accepted as probably sound, something needed to be said to show that the procedure used was thorough enough to justify acceptance in this situation. Among what Aristotle thought of as special *topoi* was the array of possible sayables that were appropriate for pursuing a certain kind of purpose, given the special type of audience charged with making judgments about a particular sort of situational problem. Searching for such appropriate sayables can be systematic, as Aristotle and other rhetorical theorists have shown.

Let me illustrate how this method of finding sayables works. The purpose of achieving expediency and avoiding inexpediency leads a political rhetor to such popular materials as what happiness is and what its constituents are, within the audience's conception of what is good. Within their conceptions of happiness and goodness lie appropriate themes and arguments for or against a particular policy.[15] Similarly, anyone wishing to praise or to blame something or someone needs sayables relevant to what is considered noble or base, including what is customarily said about various virtues and vices.[16] One seeking to accuse or to defend can draw upon conventional notions of pleasure and things pleasant as probable motivating factors for criminal actions.[17] Today, if one wants to be judged scientifically sound, one needs to refer, among other things, to one's thoroughness and accuracy.

How does a rhetor decide which rhetorical purposes should be pursued within a rhetorical situation? Cicero wrote: "The universal rule, in oratory as in life, is to consider propriety."[18] Here, "propriety" is not a matter of form but of audience adaptation. All audiences in all situations have notions of what is and is not approprite for them, where they are, making the particular judgments asked for, at a particular time. We cannot say any set of proprieties applies across all situations or is held by all audiences, but we can perceive that various communities of thought develop conventional proprieties that they then apply when judging discursive interactions among members of their particular communities. What rhetors have to do, then, is choose purposes and contents likely to be deemed pertinent and conventionally appropriate by the specifically situated thought groups that comprise the audience. Making discourse that fits conventional proprieties is not always an easy task. Propriety is a complex tapestry. It is woven of the sturdy cloth of accumulated community experience, of the community's standard practices and customs, and of both the language of everyday life and the language specific to different communities of thought. Propriety's fabric is a blend of special circumstances, age, status, and position, and the perceived nature of whatever is under discussion. "Appropriateness" in this sense is never merely a matter of fitting content to form or vice versa; it is always careful adaptation to the constraining influences of communities' partly stable and partly situationally changing principles of valuation.

When Aristotle looked for shared standards in the Athenian *polis*, he found relatively stable and widely shared norms; but he saw, too, that even those standards shifted as homogeneous Athenian citizens moved from one kind of rhetorical situation to another. Cicero saw—some would say he idealized—a communal sense of propriety in the Roman state. But he was careful to note that the situationally shared standards of the Senate were not the same as those Romans applied in the public square outside the Senate. Both Aristotle and Cicero perceived the situational variability of standards for communicative relationships. Those perceptions were important contributions to rhetorical theory.

In many communicative situations those addressed are significantly heterogeneous. A modern rhetorical theorist, Richard McKeon, addressed heterogeneity as a global problem. In the modern world we cannot assume, McKeon said, common standards for thought, language, and action.[19] Given cultural differences, attempts to communicate effectively are apt to require parties holding basically different outlooks to forge mutually acceptable normative principles.[20] They can do this only through rhetoric. Under conditions of heterogeneity, normative standards have to be created rhetorically before other rhetorical purposing can be undertaken. Once shared norms have been established, they can become the targets of further rhetoric.

Identifying or creating perceptions of shared problems among otherwise heterogeneous parties can further constrain appropriate purposing. Put in terms of Bitzer's theory of rhetorical situations: If no exigence exists within heterogeneous circumstances, a rhetor's first task is to *create* an exigence for which the rhetor's ultimate purpose can be a solution. Or, as McKeon would have put it: If heterogeneous parties fail to perceive mutual interdependence about a problem, they have no motivation for continued discussion and negotiation; therefore, rhetors must create a view of a problem as hindering the parties' interests and as requiring the participation of each party.[21] Without recognition of shared problems, heterogeneous parties will not likely forge and submit to mutually constraining normative principles for situationally appropriate discursive purposing.

The general principle is that in the presence of heterogeneity conditions of exigence and normative patterns of judgment may have to be creatively forged before rhetorical identification can be evoked. This is also applicable to rhetorical situations in which, say, both humanists and scientists are present. But even if this practice is followed, newly created exigences and norms are likely to have less importance for different ideational communities than old exigencies and norms. Scientists and humanists might forge new conditions of exigence and new norms together, but humanists adhere more strongly to those exigences and norms that bind them together *as* humanists, and scientists identify more strongly with problems and norms that bring them into association *as* scientists. Nonetheless, the purpose of making rhetoric may have to be to create shared recognition of problems and shared normative patterns of response—however weak—before undertaking to induce cooperative attitudes and actions toward other goals, beliefs, and actions.

A GENERAL THEORY OF RHETORICAL ENDS

The summative point of this chapter is that if rhetorical communication is to have success, its content must be chosen and formulated so as to seem reasonable at a particular point in discourse to a particular audience in its particular situation. Any comprehensive theory of rhetoric must explain how a person discovers what content is available and how that person selects from available sayables those most appropriate. Psychological perspectives have not given us these understandings. The question is: What general principles of purposing can guide these inventional decisions, regardless of the kind of discursive community being addressed?

A first principle of a general theory of rhetorical ends is: *All objectives, goals, ends, and aims of rhetorical communication are persuasive.* Classical rhetoricians almost uniformly agreed with this principle. Contemporary rhetoricians such as Kenneth Burke, Richard Weaver, and Chaim Perelman have contended that all rhetorical discourse is persuasive because of the inducing qualities inherent in using symbols, especially linguistic symbols and structures.[22] Minimally effective rhetorical purposing thus requires that general aims selected be reasonable, given the kind of audience being addressed and the kind of situation all are interacting within.

A second principle in a general theory of rhetorical ends is: *Ends or aims in rhetoric must be formulated with reference to normative principles that characterize a discursive community's range of situationally appropriate exigences and standard patterns of thought and judgment.* Rhetorical activity must be adjusted to principles that situationally constrain the discursive life of community members whenever they serve their community as rhetors and audiences. Rhetorical success requires rhetors to choose aims that address those kinds of exigences which recurrently concern members of the discursive community being addressed. Further, to be judged reasonable, rhetors must do this in accordance with standard patterns of thought and judgment that are communally sanctioned as appropriate by members. Thus, a rhetor's chosen aims will suggest contents that suit the norms conventionally governing relations between rhetors and audiences in the kinds of problematic situations that typically confront a discursive community, of which the immediate rhetorical situation is an instance.

The appropriateness of a rhetorical end is determined by whether it appears to advance some conventional normative principle. To take a nonclassical example, a teacher-rhetor must, first and foremost, show students where and how the content to be learned integrates with other knowledge and substantive concerns that they possess. The overarching emphasis must be on how and why the new knowledge fits into what the students already know, believe, and want. Whether the students are ready to understand the new knowledge, to be convinced of its truth, or to find pleasure in it are subsidiary considerations. Those questions do not go to the heart of the teacher's overall problem of choosing a dominant, substantive purpose. The fundamental requirement for general rhetorical purposing is that a dominant rhetorical purpose must be such that what one says will be made to intersect the substantive knowledge framework and related goals of those with whom a rhetor is to interact in a particular situation. Effective teaching inserts knowledge into respondents' existing bodies of knowledge. To say merely that one wants students "to understand" or that "address to the understanding" is the teacher's purpose is to ignore the situational context in which rhetorical interaction always takes place. Fitting pedagogical content into the students' existing knowledge sets a clear purpose and directs the rhetor's creative energies toward specific kinds of appropriate content.

Appropriateness does not have monistic meaning; it represents a plurality of meanings that correspond with the types of problematic circumstances that different communities confront. There is for every community a fundamental sense of what it means to think and to act as a member of that community. Certain norms of propriety govern the conduct of politicians and lawyers, but there are other principles that guide appropriate discourse and action among scientists and academics. Operation of these conventional norms becomes most evident when a member's words and deeds are alleged to be in violation. Thus, a scientist's legitimacy as a scientist can be attacked when his or her rhetorical goals are judged to be consonant with such extrascientific concerns as achieving celebrity, becoming a commercial success, or seeking political advantage. In contrast, a scientist who is judged as one with undeviating attention to advancing scientific knowledge is likely to be praised for possessing virtues befitting a purposive scientist. Similarly, the ethos of academicians is evaluated by specialized audiences according to how well those academicians advance the knowledge of their disciplines by publication of scholarly books and by articles published in reputable journals. Journalists' norms for judging the ethos of fellow journalists obviously are different.

A third principle of a general theory of rhetorical ends is: *A soundly conceived rhetorical purpose implicitly guides a rhetor toward sayables appropriate for particular rhetorical communication.* In the sense of discovering appropriate *topoi* and detailed content, invention can be facilitated only by identifying a rhetorical purpose that has relevance for a particular audience and situation. The end selected then has heuristic utility in revealing content suitable for advancing the end. When a rhetorical purpose is identified, ideational options come into view, enabling a rhetor to discriminate appropriate from inappropriate sayables. Thus what is facilitated is discovery of content that will seem reasonable given the rhetor's overall objective, the audience, and the kind of situation all are working within.

We arrive here at dual tests for effective rhetorical purposing. A general rhetorical purpose chosen according to the stipulations of sound rhetorical theory will identify the substantive task that is being undertaken; it will express a substantive goal. That goal will suggest what is and is not likely to be appropriate content for the task in the specific situation envisaged. Regardless of the discursive community that a rhetor intends to address, selection of a substantive rhetorical purpose implicitly directs the rhetor's attention toward useful content.

A fourth principle of general theory of rhetorical ends is: *Within heterogeneous situations, shared exigences and shared criteria of relevance have to be forged discursively before other functional ends for rhetoric can be formulated.* Every established community of thought already possesses common standards to some degree, but in heterogeneous situations shared standards will probably need to be created in order to bring the disparate communities of thought together into one rhetorical community for the immediate rhetorical situation. The rhetorical task is to build a sense of community among the otherwise different communities through (1) establishing a recognition of a shared problem or need and (2) creating common criteria of relevance that will define the range of possible rhetorical ends that can be pursued in seeking to resolve the newly created exigence(s). One should remember, however, that newly established exigences and newly created criteria of relevance are less likely to command firm and lasting commitments than the exigences and criteria to which people habitually adhere as members of their established, distinctive, homogeneous communities of thought.

The next chapter will show what principles and procedures are available for further narrowing rhetorical purposes by identifying points at issue in a rhetorical situation.

4

DECIDING WHAT THE ISSUE IS

Anyone who wants to induce attitudes or actions will have to decide two things: Attitudes and actions about *what?* and How shall I "insert" my inducement into the context of already existing thought? Those are eminently sensible questions to ask about any rhetorical situation. Ancient theorists evolved systematic ways of thinking about purposing and determining the relevance of available sayables. Their basic notions have proved useful in modern times, chiefly in treatments of argumentation; but I will argue that they have value wherever persuasion occurs. They apply within any situational context. The choice of where to "insert" claims involves a decision about what issue is at stake *now,* and the decision will be taken by an audience as reflecting the rhetor's *logical* astuteness relative to the subject and the situation.

Any situation becomes rhetorical by virtue of some exigence. The exigence, or problem, puts some matters "at issue" and makes other matters of less or no importance. Focusing on situationally important issues is essential for successful inducement because unless rhetor and audience agree in identifying the immediate problem, there can be no logically relevant meeting of minds. Traditional rhetorical theory provided a method called *stasis* analysis for locating the rhetorical task to be accomplished. *Stasis* analysis asks (and answers), "What is the *state* or *status* of the subject being discussed?" Since the history of *stasis* has been extensively dealt with by modern scholars,[1] I will focus on the systematic ways a communicator can discover what the issue is in a rhetorical situation.

WHAT IS *STASIS?*

The Greek term *stasis,* the Latin term *status,* and the English term *issue* all refer to the same phenomenon. If anything is debatable, there are predictable questions that have to be answered about the subject. Traditional rhetorical and dialectical theory asserted that discussion was likely to center on one or more of the following: (1) problems of fact ("Is it?"); (2) problems of definition ("What is it?"); (3) problems of nature or quality ("Of what sort is it?"); and (4) problems of action ("Whether action is appropriate in the given case"). Theorists differ on details about how rhetors discover the pertinent "points of adjudication," but ancients and moderns have generally agreed that these or similar questions need to be asked by composers of rhetoric who must focus what they will say and must pick and choose among all things sayable about a debatable topic.

The most thorough modern analysis of the Greek concept of *stasis* is that made by Otto Dieter. To summarize his research helps to clarify the concept itself. Dieter maintains that *stasis* is not difficult to define in English. His exposition runs thus: *stasis* and *status* come from the root STA, "to stand." The noun *stasis* had a variety of uses in ancient Greek. The stall in which horses stood was a *hippostasis.* Greeks spoke of the *stasis* of the wind, of the water, of the air, of the bowel, of politics. In the *Cratylus,* Plato explains *stasis* as the negative of the verb "go"; the opposite of walking, going, or moving, it meant standing still. According to Dieter, *stasis* and *kinesis* were accepted as contraries in Greek thought.[2]

Dieter further illuminates the meaning of *stasis* by referring to Aristotle's scientific works, primarily the *Physics.* For Dieter, the meaning of rhetorical *stasis* is best understood through analogy with the clear and precise meaning of the term as it was applied in the physical sciences.[3] Dieter says:

> It is the event which must necessarily occur in-between opposite movements of one subject in a straight line as well as in-between contrary movements of a subject on a line deflected at an angle of more than 90 degrees. It is immobility, or station, which disrupts continuity, divides motion into two movements, and separates the two from one another; it is both an end and a beginning of motion, both a stop and a start, the turning, or the transitional standing at the moment of reversal of movement, single in number, but dual in function and in definition.[4]

Dieter's explanation renders the concept particularly relevant to rhetoric. Rhetors propose changes. In their advocacies rhetors encounter "contrary motions." When rhetors encounter contrary motions, a point of collision arises between disparate arguments when a line of argument is disturbed by contrary argument. Advocates then make their stands relative to one another, at this point of *stasis* or at successive points of *stasis.*[5] This is as true today as it was in ancient times.

By asking specific questions a rhetor can locate, before actual discussion, where crucial points at issue are likely to occur. That determination will then indicate what content may be most pertinent in subsequent argument. If all rhetoric is persuasive, then every rhetorical situation will be one in which something contingent is "at issue." Ameliorating a rhetorical exigence will entail taking a "stand" or "stands" toward the problem(s) the exigence presents. In this sense, all rhetorical situations will contain points of *stasis*—issues—that require resolving rhetoric. Those rhetors who clearly and effectively address genuine points of *stasis* are more likely to resolve impasses and thereby induce favorable verdicts from audiences.

As we shall see later, the procedures for identifying *stases* are not the same for all rhetorical situations; but before addressing that matter, we need to see why it is that *stasis* points need to be searched out systematically.

WHY A *STASIS* PROCEDURE?

The *stasis* theory most relevant to rhetorical inquiry was the theory Cicero developed. Cicero's *stasis* theory was designed to direct intending speakers toward what could be said to constrain an audience's judgments about controversial subjects. It presupposed that all knowledge was subject to the contingencies of argument and was, in the final analysis, the product of human judgment.[6]

The procedure Cicero recommended guided processes of discovery in two ways. First, the procedure directed attention toward the central point(s) on which decisions would depend. Second, with the central point(s) isolated, a speaker could identify what sayables, opinions, and arguments might be appropriate to the specifically isolated issue(s). The Ciceronian scheme and the suppositions on which it rested are still eminently practical. To prepare for any rhetorical situation a rhetor must first be clear about what is and what is not at issue in the situation. Only with that settled can the rhetor sort out the logical acceptability and relevance of any possible materials that concern the subject at large. *Stasis* analysis also equips a rhetor to make clear to an audience the points on which their decisions *should* hinge.

For personal and cultural reasons Cicero focused most sharply on the special circumstances of legal pleading. He stressed that *stases* will occur at the point(s) of clash between conflicting pleas. Although somewhat narrow in itself, his analysis of legal *stases* still constitutes an excellent illustration of the general usefulness of guided inquiry into what should be said about any controversial, uncertain matter. In his youthful work, *De inventione,* Cicero said that all legal controversy involves dispute over questions of *conjecture, definition, quality,* and *procedure:*

> Every subject which contains in itself a controversy to be resolved by speech and debate involves a question about a fact, or about a definition, or about the nature of an act, or about legal processes. . . . The "issue" is the first conflict of pleas which arises from the defence or answer to our accusation, in this way: "You did it"; "I did not do it," or "I was justified in doing it." When this dispute is about a fact, the issue is said to be conjectural (*coniecturalis*), because the plea is supported by conjectures or inferences. When the issue is about definition, it is called the definitional issue, because the force of the term must be defined in words. When, however, the nature of the act is examined, the issue is said to be qualitative, because the controversy concerns the value of the act and its class or quality. But when the case depends on the circumstance that it appears that the right person does not bring the suit, or that he brings it against the wrong person, or before the wrong tribunal, or at the wrong time, under the wrong statute, or the wrong charge, or with a wrong penalty, the issue is called translative because the action seems to require a transfer to another court or alteration in the form of pleading. There will always be one of these issues applicable to every kind of case; for where none applies, there can be no controversy. Therefore it is not fitting to regard it as a case at all.[7]

The position Cicero expressed here was presumably orthodox Roman theory and doctrine, since *De inventione* is generally thought to be a record of what the youth was taught in school.[8]

The four questions of conjecture, definition, quality, and procedure exhaust the possibilities for controversy in judicial proceedings. They are equivalent in number and kind to the number and kinds of judgments that may be made in judicial controversy. To examine how these *stases* provide a method of discovery, I shall use some of Cicero's own clarifying examples. On disputes about fact, Cicero distinguishes among three types of conjectures, depending on whether the point at issue concerns the past, present, or future:

> This can be assigned to any time. For the question can be "What has been done?" e.g. "Did Ulysses kill Ajax?" and "What is being done," e.g. "Are the Fregellans friendly to the Roman people?" and what is going to occur, e.g. "If we leave

> Carthage untouched, will any harm come to the Roman State?"[9]

Cicero's discussion of the question of definition explains that disputes often involve differences about names attributed to noncontroversial facts. In such cases:

> The controversy about a definition arises when there is agreement as to the fact and the question is by what word that which has been done is to be described. In this case, there must be a dispute about definition, because there is no agreement about the essential point, not because the fact is not certain, but because the deed appears differently to different people, and for that reason different people describe it in different terms. Therefore in cases of this kind the matter must be defined in words and briefly described. For example, if a sacred article is purloined from a private house, is the act to be adjudged theft or sacrilege? For when this question is asked, it will be necessary to define both theft and sacrilege, and to show by one's own description that the act in dispute should be called by a different name from that used by the opponents.[10]

If there is agreement on fact and its verbal definition or description, the dispute will probably turn on the quality of the act, or on what kind of act it is:

> There is a question nevertheless about how important it [i.e., the act] is or of what kind, or in general about its quality, e.g., was it just or unjust, profitable or unprofitable? It includes all such cases in which there is a question about the quality of an act without any controversy about definition.[11]

Cicero's final question pertains specially to legal processes. This is the procedural or translative question. In general, this question is relevant "when there is some argument about changing or invalidating the form of procedure."[12] The point at issue might concern who ought to bring up the charge and against whom, or in what manner, or in what court, or under what law, or at what time. Cicero did not make the point, but procedural *stases* occur elsewhere than in courts; for example, in arguments over the authority of regulatory bodies.

Cicero's procedure for discovering the point(s) of *stasis* in judicial situations aimed at locating the appropriate point of focus for a plea, given what controversialists do and do not agree to and given the kind of judgment sought. Until the point or points of *stasis* are discovered and evaluated for their logical importance and priority in a rhetorical situation, a rhetor cannot know which facts and opinions will be of greatest value in meeting the case of an opponent and in swaying a judging listener's decision. Nor can a listener decide which one from among several potential issues requires focused attention at any given moment during discursive clashes. The general function of such a method for nonlegal, as well as legal, rhetoric was summarized by Quintilian:

> For every question has its basis *[status]*, since every question is based on assertion by one party and denial by another. But there are some questions which form an essential part of causes, and it is on these *that we have to express an opinion;* while others are introduced from without and are, strictly speaking, irrelevant, although they may contribute something of a subsidiary nature to the general contention. It is for this reason that there are said to be several questions in one matter of dispute. Of these questions it is often the most trivial which occupies the first place. . . . A simple cause, however, although it may be defended in various ways, cannot have more than one point in which a decision has to be given, and consequently the *basis [status]* of the cause will be *that point which the orator sees to be the most important for him to make and on which the judge sees that he must fix all his attention.* For it is on this that the cause will stand or fall.[13]

In general, *stasis* procedures lead one toward the point or points to be decided. Canvassing the relevance of the four questions is a process that progressively narrows the field of thought to a specific formulation of a question that pertains to particular circumstances of a case and situation.[14] Once a point of *stasis* is identified as one of conjecture, definition, quality, or procedure, typical topics for making arguments about that kind of issue can be perceived.

Consider how forensic orators can find arguments for prosecuting and defending if the *stasis* is what Cicero called a *status* of conjecture. A prosecutor surveys and weighs available topics for argument and selects those that are most useful to make a claim such as "In all likelihood the accused *did* an unjust deed, as charged." The prosecutor will consider whether a case can be made that the act was "caused" by the defendant's calculations of advantage or by sheer emotional impulse. The prosecutor will scrutinize the character of the accused for evidence of character qualities least admired by the community. The prosecutor will do this to see if it would be worthwhile to make the argument that the defendant is the *kind* of person that does the kind of unjust deed charged. Further, the prosecutor will consider whether such circumstances as time, place, occasion, and means afforded the accused opportunity to commit the act. Of course, the defense will survey topics to make arguments that contradict or minimize those of the prosecution: the defendant did not have cause to do the act; has admirable qualities of character, not those of one likely to do such an unjust deed; circumstances of time, place, occasion, and means make it improbable the defendant performed the act. The general pattern of surveying and evaluating sayables is much the same for each kind of question.[15]

The general procedure Cicero proposed has wide relevance. Cicero himself saw its universality. His general, as contrasted with purely legal, *stasis* questions are "Is it?" "What is it?" "Of what sort is it?" and the practical question of "whether to act."[16] These questions function as a method of discovery in philosophical and other disputes in that they point to available arguments and opinions appropriate to the issue and in that they allow one to discriminate different points of view that may influence judgment. The question about *fact* discriminates among different perspectives on issues concerning existence or nonexistence. The question of name or *definition* clarifies whether a controversy over what the existent is called is a result of semantic variation or genuine disagreement. The question of *kind* concerns the issue of what additional predicates and propositions can be said about what exists and is definable. The practical question of *action* raises the issue of how ideas guide public and private actions once the reality, its definition and kind, have been determined.[17]

WHERE CAN THE *STASIS* PROCEDURE BE APPLIED?

All theory of *stasis* implies that composers need categorizing questions that comprehensively identify what *could* be the range of grounds for judgment in a rhetorical situation. This notion applies well beyond the arena of legal pleading, as Cicero and Quintilian clearly saw. Legal *stases* are but one kind of special operative procedure deriving from general questions that are logically prior to special, situationally oriented questions. Both Cicero and Quintilian thought general *stasis* questions could be adjusted to the special circumstances of deliberative and encomiastic oratory.[18] Indeed, both believed that although special points at issue vary according to subject and situation, they all derive logically from fundamental *stasis* questions. Thus Quintilian claimed that predictable *stasis* questions arise concerning *any* subject of inquiry and discussion:

> We must therefore accept the view of the authorities followed by Cicero, to the effect that there are three things on which enquiry is made in every case: we ask *whether a thing is, what it is,* and *of what kind it is.* Nature herself imposes this upon us. For first of all there must be some subject for the question, since we cannot possibly determine *what a thing is,* or of *what kind it is,* until we have ascertained *whether it is,* and therefore the first question raised is *whether it is.* But even when it is clear that a thing *is,* it is not immediately obvious *what it is.* And when we have decided what it is, there remains the question of its *quality.* These three points once ascertained, there is no further question to ask. These heads cover both *definite* and *indefinite questions.* One or more of them is discussed in every demonstrative, deliberative or forensic theme. These heads again cover all cases in the courts, whether we regard them from the point of view of *rational* or *legal questions.* For no legal problem can be settled save by the aid of *definition, quality* and *conjecture.*[19]

Quintilian's observations illustrate that it is a rhetorical logic rather than a formal logic that determines the range of questions to be asked in rhetorical situations. The assumption of the entire system is that one needs to determine what listeners will need and expect in order to judge as they are asked to judge. One modern rhetorician has argued that a *stasis* system of guided questioning for analysis of deliberative controversy is very much needed, and he has offered a set of questions especially suited to political policy debate.

Lee Hultzén argued that four kinds of questions identify the points of controversy that can emerge during advocacy of public policy. Those issues pertain to *ill, reformability, remedy,* and *cost.*[20] His contention is that the logic of any deliberative situation requires that these considerations be agreed upon or no policy action can be agreed to. The subject of adequate medical care for the elderly can illustrate his theory.

"Is there an ill in the current system of medical care for the elderly?" Some might deny existence of an ill (if there is no ill, there is no basis for further argument). Others might contend, however, that an ill exists as the cost of health care rises above the financial resources of many elderly.

If an ill exists, a second question becomes pertinent: "Can the discrepancy betweeen rising medical costs and fixed incomes be cured or reformed?" This question is likely to be answered affirmatively. However, the aging process increases the need for high-priced medical services, and *that* problem quite obviously can *not* be reformed. Thus reformability is an issue in some cases but not in others.

If a problem *is* reformable, we come to a third question: "Will a proposed remedy cure the ill?" At this point, a variety of policies might be proposed: increasing Medicare or Social Security allowances; providing tax breaks; creating a new supplemental health care program. In fact, it was on this point of *stasis* that public discussion was chiefly focused from 1985 to 1988.

When any particular policy is considered seriously, a fourth question arises: "Will the proposed remedy cost more than it is worth?" Debate will ensue about the human and financial resources to be gained and lost by implementing the remedy. For instance, socialized medicine might be opposed as a remedy because it costs too much in terms of values and financial interests supporting the medical community even though such a system could potentially alleviate the problem for the individual citizen.

Hultzén's deliberative *stasis* procedure illustrates how the principles of *stasis* theory can be applied in order to locate the logical points of argument in situations other than those of legal pleading.[21] The fruitfulness of *stasis* theory extends even to communicative contexts traditionally depicted as "nonrhetorical." One such context is small-group problem solving.

Members of problem-solving groups do not automatically see problems in precisely the same ways, nor do they see available data in the same ways. Their perspectives on a given problem are apt to differ, and they may not share the same standards by which to recognize solutions or share knowledge of all available solutions.

Collaboration may require rhetorical discourse to create and articulate shared attitudes and knowledge. As McKeon saw, collaboration has to be rhetorically generated if it is not already present. The *stasis* problems of how to purpose and to find appropriate content for generating shared perceptions and agreements on content need to be solved. Even definitions are arguable. As Perelman and Olbrechts-Tyteca have pointed out, definitions are not only arguable, they are in themselves arguments, for to define is to make a claim against other possible understandings.[22] Where more than one person is involved, agreement on definitions must usually be rhetorically negotiated.

Phillips, Pedersen, and Wood have proposed what they call a "standard agenda" for identifying the several tasks that any problem-solving group must perform to complete its work. The agenda emphasizes what must be done rhetorically—by advocacy, discussion, and evaluation. The phases of the agenda are:

1. *Understanding the Charge.* The explanation of why the group has gathered and what it is to do.
2. *Understanding and Phrasing the Question.* The process of defining precisely what it is the group will talk about.
3. *Fact-Finding.* Discovering the nature of things, what has been happening, why it has happened, and what various people think about it.
4. *Setting Criteria and Limitations.* Specifying what a solution would look like and what it would do; separating the doable from the desirable; placing legal, moral, and economic limits on problem solving.
5. *Discovering and Selecting Solutions.* Examining as many alternatives as possible and comparing each to the criteria; finding the solution that best meets the criteria or building it out of other proposals.
6. *Preparing and Delivering the Final Report.* Putting into words the final decision of the group.[23]

The standard agenda identifies a sequence of steps that rhetorical exchanges must accomplish if solutions to problems are to be reached intelligently and reasonably. The agenda gives guidance to the purpose of rhetoric at any stage of problem solving, but it does not try to give guidance for discovery and situationally logical judgment of specific content.

There is still need for the guidance *stasis* theory provides to rhetors: categorizing questions that comprehensively identify what contents could be found by inquiry and determine what grounds for judgment exist within the communicative situation.[24] Missing guidelines are supplied by the *stasis* questions: "Is it?" "What is it?" "What is its nature or quality?" and "Is the action implied appropriate to the immediate situation?" Those questions guide the mind toward talk that will facilitate any of the tasks identified in the standard agenda. If, for example, a group's task is to discover and select from among solutions (Agenda Step 5), reference to the four *stasis* questions can remind speakers of what needs discussion and in what order. "Is there really a solution to the problem as earlier defined?" "Does possibility X, Y, or Z genuinely constitute a solution and not, say, a way of remedying some symptom of the deeper problem?" "What powers of remediation and what probable consequences are implied by this solution?" and "Is it feasible and desirable to apply this solution, given the criteria and limitations within which the group has earlier found it must work?" One thus logically "boxes" the options for contributing when solutions are being weighed.

Step 6 of the standard agenda identifies a straightforward rhetorical task of composing and presenting a rhetorical message for a special, situated audience. Having arrived at conclusions, many groups have to report their recommendations to audiences other than themselves. At this point the group's posture usually becomes that of an advising or deliberative speaker. In that case the standard *stasis* questions of deliberation can be invoked to discover what must be said to the prospective audience in its situation. Most probably the group must report whether a problem needs to be addressed. If it has concluded there is a problem, the group will need to report on the problem's reformability; if the problem is reformable, the remedy (or the remedies) the group proposes must be specified and compared to other possibilities. If the group recommends action, the costs and advantages to the audience must be examined. (If, of course, the deliberating group has concluded there is no ill, or that no feasible remedy can be found, its report will have to focus primarily on why the answer at one of these *stasis* points is negative.) In any case, if there must be a report, a problem-solving group completes its inquiry and then collectively assumes the role of a single rhetor who plans to address an audience in a rhetorical situation.

This illustration emphasizes that rhetorical *stasis* procedure aids invention wherever points of view collide in efforts to arrive at conclusions about contingent matters. When at least two minds are involved, the communicative tasks become rhetorical in the customary sense. And whether it is informal or formal rhetoric that is to be made, one of the standard sets of *stasis* questions will be helpful in determining what needs to be said and when in the course of communication.

Stasis theory also applies in informal, interpersonal communication. For example, individuals sometimes give "accounts" of behaviors and meanings.[25] Consider a momentary conflict between friends where one is thought by the other to have spoken in an improper way. One friend complains of the other's action. The accused counters with an "account" for the challenged behavior. In that moment *stasis* theory can be invoked to identify the nature of the conflict.

Only recently have rhetoricians sought to apply *stasis* doctrine in analysis of account making in informal, interpersonal communication.[26] For instance, Hauser points to Scott and Lyman's classification of accounts as "justifications" and "excuses" as designating two types of *stases* peculiar to interpersonal communication. Justifications are accounts that persons give when they accept responsibility for the act in question but deny the negative quality of the act. Forms of justification include:

1. Denial of injury (No harm was done).
2. Denial of victim (That person had it coming).
3. Condemnation of the condemners (They do worse; don't focus on me, focus on them).
4. Appeal to loyalties (The act is okay because it serves those to whom I am loyal).
5. Sad tales (My sad past justifies my present behavior).
6. Self-fulfillment (I gotta be me!).[27]

Excuses are accounts which admit the negative quality of the act, but deny full responsibility for it. Forms of excusing include:

1. Appeal to accidents (Forces beyond my control brought this about; it wasn't intended).
2. Appeal to defeasibility, invoking some mental element (I didn't know . . .; I was forced against my will).
3. Appeal to biological drives (Boys will be boys; men are like that).
4. Scapegoating (My behavior was really in response to the attitude or behavior of another—it's that person's fault).[28]

The forms of justification and excuse are logically optional ways of responding to criticism. As such they suggest lines of thought that are typically applied whenever interpersonal conflict involves the quality of an admitted action.[29]

Interpersonal conflicts are not always qualitative, of course. People often respond to charges that their actions are improper not by offering an account for the actions, but by denying they performed the actions at all. They would therefore contend they have done nothing that needs justifying or excusing. In such cases interpersonal communication should begin by focusing on a *conjectural* or factual stasis. Further, people often admit they performed an act, but deny that it was in any way improper. Among friends an insult would likely lead to charges of impropriety. In some cases one could respond plausibly with the explanation that the action was a friendly joke and not an insult. The point of interpersonal conflict would then turn on what "jokes" and "insults" mean within the context of the relationship. This ambiguity would involve *definitional stasis*. Finally,

Figure 4.1 Rhetorical Stasis Procedure
and Interpersonal Conflict

Stasis	Charge	Countercharge
1. Conjectural	"You called me a fool."	"I did not call you a fool."
2. Definitional	"This is an insult."	"It was only a joke."
3. Qualitative	"There is no excuse for insulting me."	"I am feeling so much pressure at work that I took it out on you."
4. Translative	"Our relationship is such that I have the right to judge the appropriateness of your action toward me (e.g., I am your friend and your behavior was not appropriate for friends)."	"Our relationship is such that you cannot judge the appropriateness of my action toward you (e.g., we are not friends so you cannot judge my behavior by interactions appropriate for friends)."

people can disagree about the type of interpersonal relationship in which they are participating. Actions that are proper for lovers are not always so for friends. Standards of propriety and the kinds of behavior that people anticipate shift as they move from one type of interpersonal relationship to another; accordingly, what behavior must be justified or excused also will vary. Indeed, when there is conflict about the type of interpersonal relationship one has formed with another, it is possible to respond to charges of impropriety by denying the existence of an interpersonal relationship that makes such charges legitimate. Should one person respond to the other's charge that he or she has "betrayed their friendship" with the countercharge, "We are not friends," standards for evaluating what behaviors are proper and should be anticipated become ambiguous. The interpersonal grounds on which the action could be justified or excused are no longer clear. Indeed, whether or not an "account" is needed or should be expected at all becomes ambiguous. In such cases there is *translative* or procedural *stasis* in interpersonal communication. Figure 4.1 illustrates how classifying claims identifies what is really at issue between parties.

In other instances—involving confusion rather than conflict, for example—*stasis* theory provides means of diagnosing the problem.[30] "When you said 'alien,' did you mean the same thing I meant when I said 'immigrant'?" is an inquiry into whether uncertainty is arising from a *definitional* difference or, perhaps, from some *qualitative* difference among persons. "I understand you, but I don't quite see how it relates to what we were just talking about" is a request for information about whether the other has neglected to identify the *quality* of his or her thought in particular circumstances, or has simply drifted away from the *procedural* constraints built up by previous remarks.

These illustrations show (1) that *stasis* theory provides lines of inquiry that aid in locating the psycho-logic of a rhetorical situation, and (2) that the kind of search suggested by *stasis* analysis can be fruitfully applied to forms of discourse other than legal and political debate—the arenas within which theory and procedure originally evolved. Some have argued for even wider applications of *stasis* theory.

No one has expanded applications of *stasis* theory further than Richard McKeon. McKeon extended *stasis* analysis and theory to philosophical, cultural, and ideological issues and discourse. He saw twentieth-century thought as strongly characterized by pluralism, but observed that cultures and groups were often interdependent despite perspectival diversity.[31] When interdependence creates a common sense of need to solve a problem, McKeon believed new means and ends of concerted thought and action must be forged through communication. A prerequisite to discovery of such common principles, according to McKeon, is mutual understanding of how and when distinctive perspectives differ on central questions. For this task McKeon proposed classical *stasis* theory and analysis as means of finding grounds from which new conceptions can be built.

Among McKeon's contributions to *stasis* theory was his observation that rhetorical *stasis* questions have universal scope.[32] McKeon applied the questions to foci of philosophical, cultural, and ideological controversy. This application enabled him to discriminate among different frames of reference by examining how answers to those questions vary across perspectives. McKeon thought the *stasis* questions were neutral, and thus allowed examination of diverse perspectives with minimal reduction, distortion, or bias. Further, McKeon believed that the *stasis* procedure is transformational. As those adhering to different perspectives answer the *stasis* questions, it becomes possible to transform ideas by taking them from one perspective and expressing them from the vantage points of another. One can thereby determine which ideas remain stable and acceptable across the perspectives. By distinguishing actual disagreement from semantic differences at fundamental points of *stasis,* one can create common principles, values, and perceptions. These can then serve as bases for concerting thought and action among persons and groups with otherwise diverse perspectives.[33]

The point here is not to argue for or against the universal applicability of *stasis* theory, but to note that valuable distinctions emerge when different discursive claims are subjected to *stasis* questions and procedures. Therefore, the final question to be considered here is: How does *stasis* procedure operate?

HOW DOES *STASIS* PROCEDURE OPERATE?

In developing any classification system one is confronted with problems of how each category should be formulated and related to other categories within the system. The categories chosen should be exhaustive of the subject they are alleged to classify. If I wanted to classify the "universe of things" exhaustively, I might propose the categories of animal, vegetable, and mineral for classifying all things that are. My claim could be evaluated on the basis of whether the typology encompassed all phenomena definable as "things." Categories chosen should also be discrete. Overlapping categories seldom clarify. So the categories I proposed could be judged also by how effectively they identified and defined distinct *kinds* of "things."

As a category system *stasis* theory is based on three major suppositions about the logic of argumentation: (1) in all kinds of controversies there will occur points of clash—*stasis* points—where argument can proceed no further until the points of difference are at least clarified and perhaps even resolved; (2) these points of clash or difference are of discrete and predictable kinds that can be identified by considering the reasonable requirements of fully proving one's case and by considering the points of agreement and disagreement; and (3) the kind of clash dictates the use of certain but not all kinds of content. Traditional *stasis* theory posits that the points of clash in any argument will always involve establishing issues either of *fact, definition, quality,* or *procedure.* The theory alleges that at least in legal cases these four kinds of *stases* exhaust the kinds of clash that can logically occur. It also alleges that the categories must be discrete but so related to one another that collectively these kinds of "stands" or points of clash describe the full range of issues that can arise in legal argument. If there were agreement on all four points, there could be no argument.

Cicero made the most forthright claims for the exhaustiveness of the four *stasis* questions when he said that they encompass all legal controversy: "There will always be one of these issues applicable to every kind of case; for where none applies, there can be no controversy."[34] His claim seems supported by the fact that ancient rhetoricians were unable to add consistently applicable questions to this list. The most they could do was point out that there are asystatic legal questions that are not susceptible to *stasis.* Among the *asystatic* problems or standing points that Cicero's predecessor Hermagoras identified was the problem of deficiency—there is insufficient evidence to constitute a case. There was also the problem of balance—equal distribution of evidence such that a state of balance cannot be overcome.[35] Among the asystatic stands Hermogenes later added was the despicable. For example, someone hires out his wife for improper relations and then hails into court one who refused to pay.[36] Hermagoras and Hermogenes each identified four asystatic stands. None is an instance of argumentative *clash;* hence Cicero's claim to exhaustiveness is not refuted by the writings of antiquity.

Cicero also made clear that the four *stasis* questions identify and describe the discrete kinds of legal controversy that can arise: "no issue or sub-head of an issue can have its own scope and also include the scope of another issue, because each one is studied directly by itself and in its own nature, and if another is added, the number of issues is doubled but the scope of any one issue is not increased."[37] Legal *stases* help rhetors discriminate among potential points at issue because, for example, a clash on a *conjectural stasis* differs in kind from a *definitional* or *qualitative* clash.

Hultzén's work suggested that a general, inclusive theory of rhetoric might need to offer *stasis* procedures for each of several kinds of rhetorical discourse. To do so would not, however, vitiate the basic presuppositions of *stasis* theory. If the coverage of a general *stasis* theory is the "universe of controversial questions," special *stasis* procedures can be offered to guide thinking about particular kinds of controversial questions that confront particular kinds of audiences concerned with special kinds of rhetorical aims within particular types of contexts. In principle it is possible to develop discrete *stasis* systems that exhaustively identify the points of difference that can logically arise in any substantive field. The possible points of "stand" will be determined by standard criteria for what is reasonable within that field of discourse, as I shall later show regarding scientific discourse.

If a system of *stasis* questions collectively exhausts what can be controverted in a field of rhetorical discourse and if the questions are individually discrete, that system identifies both the kinds of purposes one can have and the kinds of content that are most germane to argument in that field. It does not follow, however, that the questions in that *stasis* system identify points of controversy on which agreement can be reached, or that all argument must end if none of the *stasis* points identified is resolvable. There is a real possibility that terministic orientations of arguers will be such that the arguers will be irrevocably opposed to one another at all *stasis* points. Communists and capitalists would disagree initially about each of the questions of fact, definition, quality, and procedure that apply to any specific mix of social-political circumstances, and they would not be likely to share perceptions of social or political ills, reformability, remedies, and costs. A special value of Richard McKeon's thought is that he showed how applying the traditional *stasis* questions enables one to isolate the points of clash and to differentiate the positions being taken. Conceivably, with these clashing positions fully exposed, it would be possible to explore afresh what facts, definitions, qualities, and procedures the disputants *do* share. Communists and capitalists might, for example, hammer out agreements on international monetary policy. That, thought Richard McKeon, is a pathway toward discursively forging new common grounds that are necessary if contending parties are to resolve problems of interdependent concern. That is a valuable prospect, one opened up by *stasis* theory.

Whether one uses *stasis* theory to guide purposing and selecting content for discourse or uses the theory and system to examine the detailed nature of comprehensive differences, a basic question is, How do we know a point of *stasis* has been located? When we isolate a point of argument with a *stasis* question, and when the point is one that must be settled somehow before any further progress can be made in discussing the overall problem, we have located a point of *stasis,* a point of stoppage, between two contrary rhetorical motions. When one of these contrary motions is advanced by argument and evidence sufficient to break opposing stands, the point of *stasis* is resolved and further progress can be made. It is possible also for the point of *stasis* to become altered by common agreement to waive the issue. In any case it is the stoppage of argumentative movement that identifies a *stasis* point. The stoppage occurs because the logic of the subject and the occasion dictates that full agreement requires agreement (or waiver) on this point. In a criminal trial, for example, disagreement over whether A did or did not kill B must be resolved before further judgment about A's status can be made. If it is agreed by proof (or admission) that A did, in fact, kill B, it then becomes open to the defense to argue about the definition of the act—that it was an act of self-defense rather than murder. This option could not be logically relevant without agreement that there was, in fact, an act of killing. Likewise, in a policy dispute, judging a course of action logically requires agreement that certain reformable problems exist before it becomes relevant to consider whether policy X or Y is preferable as a reform. *Stasis* analysis, then, rests on the logic of decisions to be made and on the ways of making them. The criteria of this topical logic—a logic of *relevance*—will vary with the subject matter of argument, the kinds of judgment sought, and the situation in which that judgment is asked for.

CONCLUSION

Rhetorical *stasis* procedures require questions that canvass the range and kinds of judgments special types of situated audiences will make. They must be questions that exhaust the conditions of relevance within that situation. When those questions have been asked and answered, a rhetor has a basis for constructing discourse that is purposeful and guided rather than impulsive and potentially irrelevant. By helping rhetors identify what the important questions are, *stasis* procedure guides them toward content that is likely to be useful for addressing whatever is at issue before that special kind of audience. Moreover, making use of *stasis* procedures helps both rhetors and audiences discriminate among alternative positions on central points of dispute; this, in turn, assists them in evaluating and judging discourse.

Some have said that *stasis* theory simply exemplifies rhetoricians' tendency to overclassify communicative actions. This is a mistaken view. The question, "Is that relevant at this point?" is never absent when communications are constructed or evaluated. *Stasis* theory simply asserts that (1) kinds of controversy and problem solving are identifiable (e.g., as legal, as legislative, as philosophical), and (2) for each kind of controversy there will be *some* set of conventional test questions that measures the degree of relevancy discursive content has for the given situation. These seem empirically justifiable assumptions. Part of the training of a scientist consists of learning what the scientific tests of relevance are. Part of the training of a teacher consists, or should consist, of learning the tests by which to judge whether and where an idea is educationally relevant. Part of the training of a sociologist consists of learning by what tests facts and theories are to be judged relevant to sociology rather than, say, to physical anthropology. All of us apply tests of relevance as we compose and as we listen or read. Specialists do so more perhaps, than others. But do we test for relevance systematically or impulsively? *Stasis* doctrine identifies tests for systematic application.

5

RHETORICAL TOPICAL METHOD

At several points in the history of rhetoric, theorists have insisted that material for inclusion in discourse is not effectively found through the methods of formal logic nor through intuition. There is, it has been argued repeatedly, a topical logic which ought to methodize searches for sayables. One finds such different figures as Aristotle, Cicero, Francis Bacon, Chaim Perelman, and Richard McKeon arguing in various ways for a theory of *topical invention.* This chapter is devoted to the general theory of rhetorical topics and to considering how that general theory applies within specific fields of discourse.

GENERAL FEATURES OF TOPICAL METHOD

Especially from the eighteenth century until recently, it became customary to say the rhetorician either intuits or reasons out, by logical analysis, what needs to be said. The intuitional version of this position grew out of the belletristic tradition in literary composition and criticism and was influentially expressed by Hugh Blair in his lecture on "The Argumentative Part of a Discourse":

> What is truly solid and persuasive, must be drawn *"ex visceribus causae,"* from a thorough knowledge of the subject, and profound meditation on it. They who would direct students of Oratory to any other sources of argumentation, only delude them; and by attempting to render Rhetoric too perfect an art, they render it, in truth, a trifling and childish study.[1]

An alternative to "meditation" was offered by theorists more enamored of formal logic than of imaginative literature. They recommended that composers apply the principles of formal logic in analyzing the information associated with their subjects.[2]

As normally treated, logical analysis has consisted of learning and applying rules for testing the formal validity of deductive or syllogistic reasoning and of learning and implementing guidelines for assessing the degree of probability inductive inferences can be said to have. The focus of attention has been on *judging* deductive or inductive reasoning from premises to conclusions. There has been no primary attention to discovering or inventing the notions or evidence contained in the propositions from which one reasons.

What does this have to do with a rhetor's inventional problem of finding sayables? Deductive logic does not explain how substantive premises are found.[3] Inductive arguments are based at the outset on tacit or explicit general principles or hypotheses,[4] so we have the problem of explaining where these principles or hypotheses came from. In either case, logical analysis fails to answer the question "Once our rhetor sees what the *issues* are, how does he or she decide what to say about them?" Logical analysis helps us test validity or probability of claims, but it does not help us discover the general notions that guide reasoning and that serve as propositions when we make arguments.

Must we then rely, as Blair and others have done, on "intuition" or "genius" or unguided "meditation" to discover what to say? Certainly rhetorical invention can occur happenstantially or even, perhaps, by inspiration or revelation; but finding appropriate ideas is also sometimes systematic, as the most thorough rhetorical theorists have long recognized. The best explanation of systematic discovery of rhetorical content that Western intellectual history has offered is what is called the "topical method" of discovery.[5] I need not review the history of this idea,[6] but it is pertinent to outline the general features of the theory and procedures theorists and practitioners have found helpful in discovering what to say in a given case.

Rhetoric takes place under conditions of contingency and not those of apodictic certainty. Accordingly, the materials and patterns of rhetorical discourse can have only a plausible or probable status; but within any community of interest or knowledge there will exist sets of durable, standard, but informal headings of thought concerning the various subjects with which that community normally deals. In other words, an informal logic designates appropriate topics for discussion. Persuasiveness in any situation hinges to considerable degree on a rhetor's moving in these conventional directions as he or she develops ideas. A rhetor's business is to discover and establish claims having degrees of plausibility or probability. If he or she does so with artistry, the effort will involve exploring the *doxa,* or relevant body of opinion, for suitable materials and structures. The topical method of inquiry becomes important in this process.

The world of opinion—of humanly processed knowledge—is one in which data and conceptions are connected, associated, and dissociated. Peas are not just peas; they are *vegetables*. Atoms are not just structures; they are structures *of matter*. To inquire into what may be said about something is to inquire into the prevailing and possible modes of connection, association, and dissociation known to a culture, a society, a community of scholars, a particular audience. Fortunately or unfortunately, there is no all-inclusive cataloging system that identifies all the connection-makings humans use in thinking about and conducting their affairs; but no one familiar with a culture or other social group lacks ways to discover how a group classifies, typifies, and dissociates knowledge shared by the group. A theory of rhetorical invention, dating at least to Aristotle, focuses on this fact. That theory posits there are "classes" or "types" of thinkables and sayables, and that these can be systematically located if one canvasses what does and does not seem probable to the kind of persons one proposes to address.

Rhetorical invention, referred to in Latin as *inventio,* derives from *invenire,* "to come upon" or "to find." Implicit in the origin of the theory of *inventio,* then, was an implication of searching or questing. Searching for what? Not for an inventory of idiosyncratic connection-makings, but for sayables that could be relevant to and persuasive to the generality of persons of the type to be addressed. Aristotle explained this point clearly in his *Rhetoric:*

> A statement is persuasive and credible either because it is directly self-evident or because it appears to be proved from other statements that are so. In either case it is persuasive because there is somebody whom it persuades. But none of the arts theorize about individual cases. Medicine, for instance, does not theorize about what will help to cure Socrates or Callias, but only about what will help to cure any or all of a given class of patients: this alone is its business: individual cases are so infinitely various that no systematic knowledge of them is possible. In the same way the theory of rhetoric is concerned not with what seems probable to a given individual like Socrates or Hippias, but with what seems probable to men of a given type; and this is true of dialectic also.[7]

In order to "come upon" or to "find" what can seem probable or plausible to listeners of the type given by a rhetorical situation, one needs a strategy of inquiry, a heuristic procedure. That is what exploration of *topoi,* or topics, is. One can canvass the general classes and types of things thought and believed about one's subject and draw from that canvass the particular ideas that will probably seem persuasive about a particular point presented to one's particular audience.

According to the theory of topical invention in rhetoric, one carries out one's canvass in a manner similar to using the Library of Congress classification system. One learns the categorizing system in a general way, then explores what falls within the categories, then looks among those for ones that seem pertinent to one's subject and purposes. For example, if I want to find something to say about *forms* of argument, my general knowledge of the LC system would lead me to ask myself, "Is there likely to be any useful material under what Library of Congress classifiers tend to call 'argument,' or 'logic,' or 'debate'?" If I did not already know it from knowing the system, I would soon find out that "logic" is too broad and complex a category of ideas to be useful for my narrower concern with *forms* of argument. On the other hand, consulting the card catalog or viewer under the headings of "argument" and "debate" will certainly lead me to (and remind me of) the widely varying structures that arguments can have. By searching the "places" called "argument" and "debate," I enlarge my acquaintance with what is sayable about forms of argument. Exploring established categories of connection making could also work in a different way. If I know that materials having to do with how creatures process stimuli are classified in the Library of Congress system under the heading "cognition," I shall come more quickly than otherwise to available sayables about how humans process information. The better one knows how data are classified by those who sort out the materials, the more quickly and precisely one can discover what is there to be used.

The principles of topical investigation are not, then, peculiar to rhetorical invention, but they apply to searching for rhetorical content as readily as to other kinds of intellectual exploration.[8]

A set of headings, questions, or operations that makes up a heuristic topical method needs to be a list that (1) identifies the ways of thinking familiar to the prospective audience, and (2) triggers thought first in one direction and then another. It needs to generate the kind of thinking that Young has characterized as "neither purely conscious nor mechanical; intuition, relevant experience, and skill are necessary for effective use."[9] As levels of experience and skill change, heuristic *topoi* contained in a topical schema must be capable of amendment and revision. Thus, topical method contains flexible lists of heuristic categories; it never can be an exhaustive system of necessary and altogether sufficient rules for finding.

A heuristic topical method of invention must provide headings, questions, or operations that suggest relationships among items and thereby recognize the connection-making features of thought and also reveal new or unanticipated associations that are possible. The topical thinker must be able to "invent" either through intuiting or through observing possible relationships. Such invention can occur through a series of *transpositions*. An almost unlimited number of transpositions can take place as one surveys and applies a topical list. Any transposition will take the following logical form: View X in the perspective of topic Y. X might be a fact, term, proposition, controversial claim, or preconceived opinion or notion. Y could be an analytical term like "genus," a substantive term like "justice," a field-dependent proposition, or a field-independent argumentative strategy. X is usually particular and specific; Y (the topical heading) is usually a class or type that is general and indefinite as to detail. By bringing X into relation with Y and viewing X from that vantage point, X displays selective features. X is transformed and is given fresh meaning because Y brings to the fore special details and qualities perhaps previously unforeseen. A transposition thus allows new insights by letting unforeseen relationships come into clear view.

Transpositional thinking can be illustrated by using the general topic of *degree* or magnitude. Suppose I want to argue that acid rain is a serious environmental problem. I can use a list of *topoi*. If the list includes the *topos* of degree, that topic can suggest strategies that might be overlooked without the aid of a topical heuristic. Claims such as the following could be suggested: "Acid rain is one of the *greatest* environmental problems today." "The cost of installing industrial scrubbers as part of a comprehensive solution is *small* compared to the *great* environmental costs if the problem is not resolved." Of course, the usefulness of these lines of argument will depend on the audience addressed and the amount of supporting evidence at my disposal. The point is, however, that I can invent or remind myself of specific lines of thought about X (the acid rain problem) whenever I view it from the perspective offered by topic Y (the general topic of magnitude or degree).

Transpositional thinking guided by sets of topics can also lead us to lines of argument suited to special kinds of discourse. For example, suppose I want to praise as "heroic" the action of a particular individual who dashed into a burning building to be sure that no one was trapped inside. This desire of mine would call for development of themes customary in epideictic rhetoric. I might, therefore, decide to consult Aristotle's topical list of virtues as a heuristic. (The list identifies justice, courage, temperance, magnificence, magnanimity, liberality, gentleness, prudence, wisdom.)[10]

I consider the individual's action from the perspective offered by these topics to see whether I can discover themes that can be used to praise the specific action as the kind of act popularly thought to be "heroic." "Courage" could remind me to argue that the individual should be praised because he or she was courageous. From the vantage point offered by the topic "courage," daring qualities of the action will come to the fore while other qualities such as the place in which he or she acted will recede in importance. If my purpose were to criticize, I might seize on the topic of "prudence" and be reminded to argue that a person ought not be praised when his or her action is rash or imprudent. I would then shade the same qualities that could be displayed as praiseworthy but apply the perspective of "prudence," which makes them seem blameworthy qualities. Whether I choose to praise or blame, the pattern of transpositional thinking is the same: by viewing X (the particular action) from the vantage point offered by *Y* (the topic of "courage" or "prudence"), discussable relationships come to the fore and are available for my use in making rhetoric to praise or blame. In this case as in the others I have cited, topical development has to be logical in the senses that: (1) the direction of thought and the relational pattern being considered needs to be tested for psycho-logical, situational viability, and (2) any line of thought chosen needs to be consistent with and relevant to the standard system(s) of thinking used by the culture, society, and specific audience addressed. The tests here are not universally applicable tests as in formal logic; they are tests of *situational logicality*.

The last feature that characterizes any heuristic topical method is the provisional nature of its results. A search for sayables by using topics can help us probe for possible sayables effectively and systematically, but what we discover are *possibly* useful structures and materials, not *certainly* useful ones. They are provisional because the whole point of finding them is to check their usefulness in discourse for particular people in particular circumstances. Possibilities for making successful rhetoric are increased through mastery of heuristic topical method, but having many possibilities does not guarantee success. Once one has found possible sayables, they have to be screened for situational relevance and persuasiveness. Furthermore, it is always possible that no ideal or successful theme can be discovered. Aristotle recognized this when he set as the goal of his *Rhetoric* the thoroughly realistic aim of discovering "the means of coming as near such success as the circumstances of each particular case allow."[11] There is not much to say in answer to "Was there an act of violence?" if twenty witnesses saw Jones aim and fire at Smith while Smith sat in the park reading a book!

As is implied by what I have already said, a rhetor ought to select his or her ideas and arguments while engaging in at least imaginary dialogue with the audience to be addressed. Here, reflecting on the results of *stasis* analysis can be of great help. If a discovered idea or argument cannot bear on the point that will be at issue in one's envisioned rhetorical situation, the idea's secondary importance or complete irrelevance will immediately come into view. Additionally, a rhetor will need to ask specific questions such as the following about topically discovered ideas: Will the particular audience object to or question the discovered idea? Will they meet it with enthusiasm, apathy, or doubt? Are there ways of presenting the idea so as to take advantage of, or redirect, the particular audience's expected responses? The object of the rhetoric to be made is to induce attitudes and actions; therefore a topical thinker needs constantly to weigh each discovered idea for its relevance to the *stasis* point(s) requiring address and to the special interests and knowledge of the particular audience.

Also, there will be need for evidence and other materials that lend substance and credibility to one's claim. "Acid rain is one of the greatest environmental problems today" and "Individual X is praiseworthy for having the courage to run into the burning building" are not likely to be very convincing as claims without data emphasizing the gravity of the acid rain problem or the motives for X's actions. Without evidential support that is relevant, claims advanced are likely to seem mere commonplaces or clichés. As is so often the case, Aristotle's comment is incisive on this point:

> Whether our argument concerns public affairs or some other subject, we must know some, if not all, of the facts about the subject on which we are to speak and argue. Otherwise we can have no materials out of which to construct arguments. . . . This is the only way in which any one ever proves anything, whether his arguments are strictly cogent or not: not all facts can form his basis, but only those that bear on the matter in hand: nor, plainly, can proof be effected otherwise by means of the speech. Consequently . . . we must first of all have by us a selection of arguments about questions that may arise and are suitable for us to handle; and then we must try to think out arguments of the same type for special needs as they emerge;

> not vaguely and indefinitely, but by keeping our eyes on the actual facts of the subject we have to speak on, and gathering in as many of them as we can that bear closely upon it: for the more actual facts we have at our command, the more easily we prove our case; and the more closely they bear on the subject, the more they will seem to belong to that speech only, instead of being commonplaces.[12]

The topical manner of inventional searching, whether for lines of thought or evidentiary support, has gained new adherents since World War II. It has been recommended and fruitfully experimented with as a way of searching out probable, plausible, or possible materials and patterns for treating salient issues. We may conclude generally that the productive character of topical procedures rests on the following features, at least. A topic is a heading, a collective concept; it is thereby suggestive of subordinate particulars or subpatterns. To think of a topic is to exercise both intuitive and rational thought about whatever is relevant to that topic. As these thoughts arise, their potential usefulness for discussion of a given issue can be reviewed; thus, discovery and evaluation of sayables occur.

As both ancients and moderns have shown, topics pertinent to general or specific subject matters can be arrayed as lists, to be checked through as cues during the process of inventing what to say. For example, "size," "time," and "form" suggest features that are logically discussable with respect to virtually anything. Some other topics reflect the special logics of particular types of enterprises but are widely applicable within communities that engage in that type of activity. For example, "specific gravity," "molecular structure," and "toxicity" are among the special topics that can suggest pertinent, discussable matters to anyone who deals with the chemistry of any phenomena. Special topics of this sort are useful across specializations and professions. The special chemical considerations just named are potentially useful to a naturalist discussing the characteristics of plant life, a civil engineer considering waste disposal, almost any environmentalist, and, of course, any professional chemist. As I shall show in later chapters, there are, in fact and in use, a number of special topics that characterize thought about any and all sciences.

The formal character of topical thinking deserves repeating: it is viewing X in the perspective of Y, to reveal provisional aspects of X that might otherwise be overlooked. It is a mode of thinking that everyone who uses a thesaurus follows, and it is one that artists and artisans from poets to auto mechanics use constantly although probably unconsciously. Modern rhetorical theorists have tended to undervalue the heuristic worth of this ubiquitous, associative way of thinking, but there are signs of restored recognition of its value.

A series of experiments in the 1970s showed clearly that thinking about the content of impending rhetoric is stimulated and guided systematically by using sets of topics that have the general features just discussed.[13] On the evidence of these experiments, no one set of topics *(topoi)* has this advantage exclusively. A variety of topical schemes has been experimented with, ranging from systems derived from Aristotle to the category system of Peter Roget's *Thesaurus of English Words and Phrases,* or to systems constructed from subjects' own habitual ways of analyzing things. All yielded discussable possibilities more comprehensive than could be located by subjects left to their own unguided ways of conjuring up sayables. Traditional rhetorical theory proposed that useful sets of rhetorical *topoi* ought to be divided into at least two classes or levels: general or common *topoi* that guide the mind toward themes one is likely to use in any discourse on any subject (e.g., existence, degree, size) and special *topoi* that guide the mind toward themes peculiar to particular subjects and purposes (e.g., mass and tensil strength in engineering and architecture).

GENERAL AND SPECIAL *TOPOI*

The notion of general and special *topoi* was first broached by Aristotle. His discussion of the matter is confusing on some points,[14] but the distinction between general and specific topics remains useful. One knows that preachers, lawyers, dentists, theater managers, and everyone else have occasion to discuss such general themes as existence, degree, and size in their general *and* in their technical rhetoric. On the other hand, existence of the soul is a topic of discussion for the preacher but seldom for lawyers, dentists, and theater managers. Legal precedent is almost always a pertinent topic for lawyers but seldom for the others. The theme of gum disease will be peculiar to dentists, as the topic of blocked stage movements will be peculiar to the discourse of stage managers. These are recurring *topoi* of discourse that derive from the special, analytical principles used by distinct occupational groups, and they may profitably be thought of as special *topoi* that reflect field-dependent logics. Modern theory and experimentation have given little attention to how such field-dependent arrays of *topoi* contribute to rhetorical creativity. Research has focused on the values of general topics, but in principle at least it appears that special topics serve people working within a given field just as general *topoi* have been shown to enhance perception of what is sayable on general matters. It is therefore worth pursuing the theory of special topics further.

Grimaldi interprets special or particular *topoi* as "the sources to which one has recourse to develop an understanding and thorough knowledge of the subject."[15] A rhetor turns to these special places to find propositions peculiar to a particular subject. By using propositions about these topics as premises, a rhetor can make special lines of argument. As I have illustrated, however, which special topics are most useful varies across substantive fields of discourse and across situations.[16]

When considering special rhetorical topics, we must recognize that one will not become an expert on a subject through study of even a thorough list of its special rhetorical *topoi.* A rhetor may refer to the topic of metrical scansion in talking about poetry, but understanding the significance of the scansion requires a body of specialized literary information about poetic form that familiarity with rhetorical topics per se can never give. This illustrates why Aristotle thought one has to become more and more of a specialist rather than rhetorician as one moves into specialized lines of thought and argument.[17] Knowledge of how to make rhetoric about a speciality is not in itself knowledge of the technical specialty one tries to discuss, but the more technical one becomes, the more special knowledge is required. Another way of saying this is that the more field-dependent a topic is, the closer it comes to being a distinctive principle of the field and not a theme of general communication. For example, in the field of internal combustion engineering, compression ratios constitutes a special *topos* of major importance, but whatever deserves to be said about compression ratios derives from special knowledge of mechanics, not from general knowledge of what can persuade *across fields;* the latter is the purview of general rhetorical theory.

It does not follow that highly specialized, technical *topoi* are irrelevant to creating influential rhetoric. Within the universe of political discussion, special arguments about the destructive power of weapons and about management of the governmental budget will be necessary. Political speakers therefore need to know the special *topoi* and arguments relative to ways and means and war and peace. These *topoi* can be quite technical, intersecting those accountants use when discussing budgetary principles and those strategic defense planners use when assessing the strengths and weaknesses of available weapons systems. Of course, political speakers also have special *topoi* peculiar to their own trade. They need to persuade diverse audiences about policy matters. They will therefore turn to special *topoi* that yield themes and opinions concerning such matters as what a political community believes is desirable and what conduces to its political happiness. Furthermore, political speakers will also need general *topoi* such as "degree" to make arguments claiming, say, that a particular weapons system or budgetary policy will yield more communal goods than will alternatives. All rhetors use general *topoi,* regardless of specialty or the degree of diversity among their audiences. Special topics identify clusters of ideas and ways of thinking about data peculiar to a particular field of endeavor, but whoever is to create rhetoric relative to that field also needs to have general *topoi* to guide his or her search for what is sayable about special subjects.

The distinction between general and special topics is thus of practical value. General topics indicate themes for discussion about any matter within any situation and before any audience. Special rhetorical topics are peculiar to special fields or areas of investigation, special kinds of situations, and special types of audiences. The materials they identify may or may not be usefully discussed outside the relevant field; they will certainly be discussable at some times within the field.

Special topics will be more general than specific directions for investigation within a given area of specialization or than items of data themselves. For example, within classical scholarship "sophistic" is a special topic, but the concepts toward which it will direct a classicist's mind may be matters for discussion in classical research or problematic matters inviting research prior to suasory discussion. The *topos* sophistic reminds a specialist that the characteristics of pre-Socratic Sophists and of teachers associated with the so-called Second Sophistic are widely recognized themes for profitable discussion and argument within the field of classics; but a specialist would also recognize that, say, sophistic influence on St. Augustine's theory of Christian preaching remains at this point much more a topic for technical research than for even professional exposition. Similarly, the *topos* "toxicity" can expose the presence or absence of data about a particular chemical, but this technical information may or may not be relevant to a matter of professional discussion. In any case, while the presence or absence of technical information may be brought into view by thinking of a special rhetorical topic, such information may or may not be useful in rhetoric for a given occasion. Further, the *topoi* useful in rhetorical discovery do not systematically expose to view all that is known or needs to be known in order to fill out a body of technical knowledge.

Topical method supplies a fertile vantage point from which to understand the rhetoric that goes on within any substantive field of discourse.[18] For this and other reasons several scholars have argued that topical method constitutes an important means of addressing problems and achieving objectives that specially concern (or should concern) their respective disciplines.[19] Some have even begun to discuss the rhetoric of their own academic fields of inquiry from a topical point of view.[20] There is good reason for this. As much as any politician, lawyer, or cleric, successful academics concerned with producing and professing knowledge of substantive fields also require mastery of the rhetorician's craft of symbolic inducement in general and of persuasive argument in particular.

Although Robert Nisbet nowhere uses the terms "rhetoric" or "*topoi*," his provocative book *Sociology as an Art Form* is the product of a mind thinking topically about the special logic of a substantive academic field. Nisbet's study yields a compendium of special *topoi* which designate the coverage and boundaries of reasonable thought and discussion about sociological matters. A pivotal concept guiding his study is the "theme." In Nisbet's analysis themes, like *topoi*, help users provide "an ordering of experience" and help put observation in "a special focus."[21] A theme, regardless of its specific field, art, or science, "has a generality and also an evocative power sufficient to direct the motivations and energies of countless individuals each of whom may be unaware that what he is doing in laboratory, studio, or study takes on significance as part of a prevailing and also coercive style or theme."[22]

Nisbet works like a rhetorician when he delineates the compelling themes for making sociological discourse. He specifies "master themes" (topics) common to all of the social sciences and all social philosophy: individual, order, freedom, and change.[23] Special themes (or special *topoi*) of sociology are based on the master themes. These include community, authority, status, the sacred, and alienation. These concepts or lines of thought are peculiar to a particular kind of discourse: They specify the range of reasonable thought and discussion in sociology. These themes, Nisbet explains, "may be seen as forming in their combination the essential structure on which have been built the diverse concepts and theories of modern sociology."[24] Additional lists of special topics for doing sociology are ordered by Nisbet under the headings of sociological landscapes, sociological portraits, and (for lack of better headings) what might be called "social motion" and the "malaise of progress."

Nisbet's study identifies special topics and catalogs them so they express the structure of the substantive field of sociological discourse. This topical approach can be extended in principle to any art or science.

If special topics used in a particular field of discourse can be brought to conscious awareness, it must be granted that general topics apply as well. General topics, by their nature, guide the mind toward ideas and patterns of inference that are likely to come up in any discourse in any situation. They must, then, occur in all fields. Although lists of general *topoi* vary according to the terminologies and levels of analysis preferred by commentators, it is not possible to engage in sustained and coherent discourse without using the kinds of ideas or patterns that any list of general *topoi* will designate. Ideas like cause and size and degree will always be needed and used. It is therefore not surprising that rhetorically oriented scholars have identified and made explicit lists of general *topoi* that are used in special fields of discourse. Among the themes of argument McCloskey pulls from analysis of the rhetoric of economics are general topics such as authoritative appeal, definition, analogy, and symmetry.[25] Kallendorf and Kallendorf articulate a set of general topics for corporate speechwriting. They organize these under the headings of definition, comparison, and consequences.[26] Similar studies can be made of discourse in any field; such is the nature of general topics. What these topics are *called* is not fundamental; the fundamental point is that we cannot talk without alluding to such matters as comparisons, causes, consequences, degrees, and the like.

Classical rhetoricians provided wise counsel when they responded to the problem of finding what to say in rhetorical communication with organized lists of special and general *topoi*. These two kinds of *topoi* illustrate two levels or directions of thinking that are required when preparing rhetorical presentations systematically. One must talk as others talk, and one must talk about specific subjects. The utility of this approach for solving the problem of "discovery" is shown by the inclusion of similar sets of special and general topics in the best contemporary public speaking and composition texts.[27] Contemporary scholarship in general shows that studying the topics of academic or scientific fields of discourse illuminates the structures of thought peculiar to those fields. Given that some kinds of general topics apply to discourse in all fields, the more challenging task is to extract the special topics peculiar to various substantive fields. My focus in later chapters is the special topics of the rhetoric of science. This necessitates more detailed explanation of the characteristic features of special topics, or field-dependent *topoi*.

FIELD-DEPENDENT *TOPOI*

A characteristic feature of a genuine rhetorical situation is the presence of the ambiguous, the indeterminate, or the uncertain. There is a potentiality for difference, division, conflict, or faction, and this makes unifying rhetoric necessary. If a rhetor can focus thought on alignments, classifications, and relationships of ideas with which the audience is comfortable, that rhetor will also have suppressed otherwise divisive features of the rhetorical situation. The resources to build unifying, perspectival perceptions of situations will vary from situation to situation, but where rhetors and audiences perceive themselves sharing a "field" of interest or a "community" of experience, the means of inducing unifying adherence become more predictable. Lines of thought that can rouse and intensify a sense of shared identity can be known beforehand. Whatever enhances the dignity of the law draws lawyers together as lawyers, and whatever measures with accuracy draws unifying respect from scientists. In theory of rhetoric it is the function of systems of special *topoi* to collect, under superordinate headings, interrelated sets of sayables that have these unifying potentialities for a special kind of audience.

To illustrate this and other important conceptual features of special topics I draw on an article that has not received the attention it deserves. In 1962 Steele and Redding developed an inventory of shared values germane to American culture. They argued that this inventory identified sets of premises for the construction of persuasive enthymemes.[28] As "concepts of the good" embraced by American audiences, the inventoried items influenced and constrained how American audiences were predisposed to perceive, comprehend, and react to any rhetor's assertions and claims.[29] In short, the themes Steele and Redding identified manifest all the features of special rhetorical places for finding specific arguments about American experience. According to Steele and Redding, the special rhetorical *topoi* suited for making political discourse in the United States are:

1. Puritan and Pioneer Morality
2. The Value of the Individual
3. Achievement and Success
4. Change and Progress
5. Ethical Equality
6. Equality of Opportunity
7. Effort and Optimism
8. Efficiency, Practicality, and Pragmatism

9. Rejection of Authority
10. Science and Secular Rationality
11. Sociality
12. Material Comfort
13. Quantification
14. External Conformity
15. Humor
16. Generosity and Considerateness
17. Patriotism

These *topoi* endure in the American universe of political discourse. They designate themes Richard Nixon used in 1952 when making his expense-fund speech—a speech that was enormously effective with the American public.[30] The same *topoi* were still in use thirty-two years later in the unifying lines of thought Republican campaigners and spokespersons used while exhorting the citizenry to reelect President Ronald Reagan. These *topoi* also indexed some of the themes former Vice-President Walter Mondale and his supporters tried to make compelling during their unsuccessful bid for the presidency in 1984. As this is being written, the 1988 presidential campaign has just closed. It was a campaign widely called dull, negative, and uninspiring, yet both Vice-President Bush's and Governor Dukakis's rhetoric prominently drew on themes such as change and progress, the value of the individual, achievement and success, and sociality. Indeed, the charge that Dukakis was an "ice man" led him to special efforts to display his sense of humor, and the issue of which candidate was or was not "in the mainstream of American politics" was argued repeatedly.

Every field of discourse contains shared preconceptions or patterns of thought that are so compelling that participants within each field do not have to justify the fact that they use them. Such common themes are located by referring to systems of special rhetorical *topoi* designating the characteristic lines of identification that make a discursive community possible. Without clusters of commonplace ideas and familiar patterns of thinking, a field of discourse would not be discernible; nothing would distinguish the range and forms of one field's discourse from discourse in other fields. Later I shall show that scientists have a set of approved ways of thinking which cluster around a set of special *topoi*. Such *topoi* identify the themes that express and extend identification among members of the scientific community.

A useful topic's specific meanings are multiple; a useful special *topos* is always an ambivalent term or phrase. Politicians may be committed to each and every one of the *topoi* identified by Steele and Redding. Such terms and phrases guide rhetorical choices, but do not determine those choices. It is possible for opponents in a political dispute to embrace all of Steele and Redding's *topoi* and nevertheless argue toward quite different conclusions. This does not mean that they are irrational. It is possible because each *topos* is imprecise and superordinate to all views on the designated theme. There lies the heuristic richness of the topic as a *topos*. The political *topoi* of equality of opportunity, the value of the individual, achievement and success, sociality, and so on can elicit identification from a whole spectrum of political thinkers representing a number of ideological divisions. Each *topos*, as a category of diverse arguments, transcends division. Different factions can infuse abstract *topoi* with meanings peculiar to their own orientations, and thus discover workable themes by thinking about the *same* topic that the opponent thinks about. Conflict will turn on whether one is for or against a specific application of a *topos*, or whether one is for or against a comparative ordering of the *topoi* according to their alleged significance. Consider some examples.

Advocates of more stringent environmental controls may turn to the *topos* of the value of the individual. Calls for more stringent controls can be justified by arguing that they safeguard the individual citizen from improper waste disposal by giant chemical corporations. On the other hand, chemical companies can find arguments by consulting the same *topos*. They can, for example, adopt the persona of an individual burdened by the weight of government restrictions and argue against additional safeguards as being unduly constraining. Each political actor identifies with the *topos* in the abstract, and then by moving in thought from the abstract to concrete applications, each discovers an array of specific arguments supportive of its overall position.

Conflict can also result over the hierarchical arrangement of *topoi*. For example, a spokesperson for the American Civil Liberties Union may argue from the *topos* of the value of the individual that police powers ought to be restricted in search and seizure operations. A police spokesperson may discover an argument within another, allegedly more important, *topos* and counterargue that a relaxation of restrictions will make law enforcement more efficient. That argument is associated with the *topos* of efficiency, practicality, and pragmatism. The dispute will now turn on which *topos* is the more important.

Topoi point to what may be said logically on any side of a controversy. What the *topos* suggests is a perspective from which or within which to argue. A *topos* specifies a vantage point from which to view issues, ideas, and facts.

It is important to notice that while specific issues, ideas, and facts may change over time and across rhetorical situations, the perspectives or vantage points from which people choose to view their circumstances do not change rapidly. It is this that gives rhetorical *topoi* their vitality and durability as aids to rhetorical invention. Special *topoi* do change, but the process usually is gradual. New rhetorical *topoi* emerge as people discover fertile new relationships that transform understandings of old ideas.[31] It took some time following the appearance of *Origin of Species* in 1859 before the *topos* "evolution" entered the professional thinking and parlance of historians, sociologists, and others, for example. Fresh analogies and metaphors can suggest new modes of classification and relationship within a field of discourse. Kuhn's "paradigm shift" and Chomsky's "deep structure" were such metaphors in history of science and in linguistics, but they gradually came to identify patterns of thinking in other fields as well. Extended analogies from other fields sometimes yield reorientations regarding a given field's problems and how they ought to be resolved. The theories called social Darwinism reflected this kind of development in nineteenth-century social science. Fresh insights may call for new special *topoi* to characterize the new perspectives. Following the emergence of impressionism in art in the nineteenth century, the *topos* "mood" took on new importance in criticism of music and painting. The principles of thought that new *topoi* suggest gradually gain vitality and durability, sometimes within a single community and sometimes within many; the principles of evolutionary development enter into the thinking of virtually all intellectual communities today.

On the other hand, once vital and important *topoi* can wither or decline into stereotypes or clichés possessing little evocative power.[32] A rhetorical *topos* runs its full course when its initial power to suggest provocative new relationships of similitude and difference develops into a concept that holds together a seemingly enduring network of ideas and relationships and then devolves to a stage of irrelevant or stagnant ideas. The Aristotelian special *topos* "magnificence," suggesting an important theme of praise for a person, has had this fate in the West. Despite potentiality for change, however, communal *topoi* are strikingly durable. They can change only if the community that bred them changes markedly in its patterns of collective thought.

To this point I have presented the general theory of rhetorical creativity that has evolved from antiquity to the present. It is a portrayal of persuasive creativity that allegedly applies to any genre of inducing discourse. It specifies the kinds of ends that are rhetorically possible. It proposes procedures for discovering what end or ends ought to be pursued in a given rhetorical situation. These procedures centrally involve discovering what is at issue in a rhetorical situation. The concepts of *stases* and their possible forms provide means of analyzing and responding to different sorts of stands or issues. The theory I have summarized also posits the social existence of *topoi,* or topics that index potentially discussable themes both generally and specifically.

What inherited and modern rhetorical theory offer concerning the discovery of what needs to be said is a system of heuristics by means of which communicators can systematically and logically discover instrumentalities for adapting inducible ends within situations that are open to change through the mediation of practical discourse. In the remainder of this book I shall inquire into whether similar heuristics apply to the informal logic of scientific discourse. If so, I shall try to set forth the character of a theory of rhetorical invention derived from the practice of specifically scientific communication.

Part II

RHETORICAL INVENTION IN SCIENCE

6

RHETORICAL DIMENSIONS IN SCIENTIFIC DISCOURSE

A general theory of rhetoric specifies what is unique to rhetorical thinking compared to, for instance, formally logical thinking. In this chapter I will begin to develop an adaptation of that theory that applies specifically to scientific discourse.

To isolate rhetorical dimensions of scientific discourse I need a fair and accurate description of the processes and qualities of discourse that is counted as "scientific." Kuhn supplies that description, at least in part. Whether his conception of science as an enterprise is philosophically adequate is not crucial to my purposes;[1] whether he fairly describes what scientists *do* in discoursing with one another is crucial. Some historians of science quarrel with how Kuhn treats some of the historical instances he uses to support and amplify his key conceptual claims,[2] but few challenge the general, descriptive adequacy of his analysis of science as a discursive activity, whatever else science may be. It is therefore justifiable to enter an analysis of scientific discourse, as discourse, through Kuhn's description.

The first point of importance about Kuhn's perspective on science is that *persuasion* is a prominent feature of discourse that brings about scientific maintenance and change. There is a nascent rhetorical perspective in Kuhn's analysis of science, as a good many rhetoricians and some commentators on science have recognized. However, neither Kuhn himself nor those who have seen the rhetorical dimensions of his analysis have fully explored and developed those rhetorical dimensions. For that reason I shall undertake to show how scientific activity is and must be rhetorical in the ways identified in preceding chapters.

SCIENTIFIC DISCOURSE IS SYMBOLIC INDUCEMENT

Let us begin with Kuhn's concept of the paradigm. The term "paradigm" has been used so extensively since publication of *The Structure of Scientific Revolutions* it has been demoted to an ambiguous academic cliché. Kuhn's initial formulation of this idea was at least partly responsible. In an otherwise sympathetic commentary on Kuhn's work Margaret

Masterman demonstrated at least twenty-one different senses of the term "paradigm" as it was used by Kuhn.[3] In response, Kuhn explained that most ambiguities in the concept's meaning are due to stylistic inconsistencies, and that all of its uses are reducible to two distinguishable but overlapping meanings. First, he said, there is a sociological sense to the term. It designates components of consensus that hold a scientific group together. Second, the term denotes one of those components: the concrete, exemplary problem-solutions that constrain and guide the research of a scientific group. As he explained,

> On the one hand, it [paradigm] stands for the entire constellation of beliefs, values, techniques, and so on shared by the members of a given community. On the other, it denotes one sort of element in that constellation, the concrete puzzle-solutions which, employed as models or examples, can replace explicit rules as a basis for the solution of the remaining puzzles of normal science.[4]

Kuhn's sociological view of science focuses on the components of consensus that function in a scientific community to hold the group together.[5] This "constellation of group commitments" accounts for the relative fullness of professional communication and the relative unanimity of professional judgments within the group. To clarify and extend the sociological meaning of "paradigm," Kuhn introduced the idea of a disciplinary matrix:

> I suggest "disciplinary matrix": "disciplinary" because it refers to the common possession of the practitioners of a particular discipline; "matrix" because it is composed of ordered elements of various sorts each requiring further specification. All or most of the objects of group commitment that my original text makes paradigms, parts of paradigms, or paradigmatic are constituents of the disciplinary matrix, and as such they form a whole and function together.[6]

Kuhn did not provide exhaustive coverage of the constituents of the disciplinary matrix, but he did identify several kinds that, if present, promote consensus in thought and action. Included among these constituents are symbolic generalizations, models, values, and exemplars. Symbolic generalizations are formal or formalizable expressions that members of a scientific community can employ without question. Expressions such as "f = ma," "action equals reaction," or "all cells come from cells" are used without need for justification. Shared commitment to such symbolic generalizations is required if the powerful techniques of logic and mathematics are to be applied routinely in the community's work.[7]

Research models comprise another component of the disciplinary matrix. Models supply a group with acceptable analogies and metaphors. Some are used heuristically while others have the force of a metaphysical commitment. Heuristic models serve guiding or suggestive functions in a community's research. For instance, consider such statements as "The electric circuit may fruitfully be regarded as a steady-state hydrodynamic system," or "Gas behaves like a collection of microscopic billiard balls in random motion." A heuristic model is not considered isomorphic with the phenomena being investigated. Metaphysical models are, however, used to express commitment to such direct identifications. Examples include such statements as "The heat of a body *is* the kinetic energy of its constituent particles," or "All perceptible phenomena are due to the motion and interaction of qualitatively neutral atoms in the void."[8]

Kuhn distinguished values from symbolic generalizations and models.[9] As he suggests, scientists otherwise separated by differing foci of specialized research and disciplinary interests nevertheless attain a sense of community with all other scientists through adherence to widely shared values. Thus, commonly held values may be used across several sciences and within the numerous specialty fields to judge theories as wholes. Valuable scientific theories generate solvable puzzles, and possess virtues of simplicity, accuracy, internal consistency, and external consistency with other theories currently applied to related domains of phenomena. Of course, how these values are applied by particular scientific groups can vary. So too can application of values to particular claims advanced by members of a single scientific group. Nonetheless, shared values remain a feature of communal identification.

An exemplar is an element of the disciplinary matrix that denotes a concrete solution to a problem which has gained acceptance throughout a scientific field as a model of appropriate investigative procedure. A field's exemplars thus prescribe the range of acceptable problems and permissable solutions for its practitioners.[10] They also constitute the core of scientific pedagogy. Kuhn identified a special function of exemplars that makes them an essential resource in scientific pedagogy: Exemplars provide concrete empirical content for the language used to express concepts, theories, and laws learned by students. As Barnes explains it,

> Without paradigms [exemplars], there can be no understanding at all of the proper usage of scientific terms; how such terms apply cannot be deduced from their abstract presenta-

> tion in laws and theories. . . . Paradigms are to concepts like "mass" and "force" what concrete instances are to concepts like "cat" or "shirt."[11]

Students learn to solve problems by seeing resemblances between known exemplary problem-solutions and the specific exercises they are asked to work on. Learning to see that a new problem is like some problem already encountered and solved is the very basis of scientific training. Well-trained scientists are prepared to grasp similarity relationships between known problem-solutions and unsolved problems and thus are ready to take their places as highly competent "puzzle-solvers" within a research community.[12]

All disciplinary groups have sociological paradigms or disciplinary matrices, but only mature sciences have concrete exemplary paradigms that are shared collectively. It is the exemplary paradigm that makes the practice of paradigmatic science more constraining than the practice of a nonscience or immature science. As two sociologists put it,

> The function . . . of an exemplar is to permit a way of seeing one's subject matter *on a concrete level,* thereby allowing *puzzle solving* to take place. . . . For a discipline to be a science it must engage in puzzle-solving activity; but puzzle-solving can only be carried out if a community shares concrete puzzle solutions, or exemplars.[13]

If a discipline, say sociology, is ever to attain the status of a mature science, it must first develop exemplary achievements in substantive fields such as political socialization and not rely on the triumph of general theoretical orientations such as functionalism or symbolic interactionism.[14] On Kuhn's terms, no discipline can attain full scientific status unless it can include shared exemplary paradigms within its disciplinary matrix.

It is also the exemplary paradigm that promotes firm consensus among a scientific community's practitioners. In the social sciences "master theories" like Freudian psychology, functionalist sociology, and Keynesian economics supply guidance for social scientific practice and inquiry, but this guidance is not exclusory; there are "reasonable" alternatives that stimulate continual reassessment of fundamental theoretical and methodological questions and assumptions. This is not the case in mature sciences. Gutting put the point thus:

> Various social scientific super-theories have, during certain periods, been in some sense widely accepted, at least among certain subgroups of researchers. But the sign of Kuhnian

> consensus is not just some sort of general endorsement of a super-theory but an acceptance that is so strong it eliminates the need for further discussion of foundational questions about the subject-matter and methodology of the disciplines and enables the discipline to devote most of its energy to puzzle-solving. A consensus that does not have this character will not be sufficient to sustain the practice of Kuhnian normal science; and it is this practice that is the mark of a mature science.[15]

Kuhn's concept of disciplinary matrices clarifies by describing constitutive features of scientific consensus. I wish to stress, as Kuhn does not, that these constitutive features exist as bases for rhetorical activity. This is implied when Kuhn focuses on persuasion as the operative feature that brings about change of scientific paradigms. Kuhn's view of persuasion is, however, conceptually underdeveloped. Any discourse is rhetorical insofar as it involves selective use of symbols to induce cooperative actions and attitudes regarding particular orientations for attaching meanings to situations. Scientific discourse is rhetorical precisely for this reason: It involves selective use of symbols to induce cooperative acts and attitudes regarding a mediating symbolic orientation. Selectivity involved in making and evaluating scientific discourse indicates the rhetorical dimensions of that discourse. As Campbell put it, any scientific statement is a *partial* story, regardless of its accuracy or precision, and the partial story always involves at least subtle advocacy.[16] Further, any coherent terministic claims define an orientation toward phenomena or provide a way of viewing phenomena that reveals features and relationships one might otherwise overlook. For discourse to be meaningful, selected symbols must induce cooperation with a coherent orientation, and this is as true of scientific meaning as of any other kind of meaning. Generally speaking, a terministic orientation does not require rejection of other ways of "seeing," but a paradigmatic orientation is always insistent and in a mature science is usually exclusory. Paradigmatic discourse is, therefore, the most tightly constrained kind of rhetorical discourse in science.

The paradigm concept emphasizes the importance of selectivity in scientific activity. A paradigm constrains the vision of those who accept it; it suggests what is to be anticipated in respect to situated phenomena. By restricting a scientist's vision of a research field, a paradigm directs attention to certain problems and possible means of solution; to acceptable facts and phenomena suitable for scientific investigation; to meaningful applications of terms, labels, concepts; and to standardized laboratory procedures and relevant equipment. Further, groups of scientists will share a vision of what constitutes appropriate scientific activity. That vision will have been previously negotiated in formation of paradigmatic consensus. Some of the vision's elements are identified by Kuhn's notion of the disciplinary matrix. In sum, selectivity is involved in the very acts of learning, choosing, and using a paradigm because potential alternatives are excluded and a special claim is made on behalf of the paradigm chosen.

Commentators other than Kuhn have also recognized that decision making, negotiation, and judgment permeate the activities of research scientists. Knorr-Cetina argues that scientific inquiry is not a process of careful description aimed at securing absolute truth, but instead is a *constructive* activity involving several series of selections aimed at "manufacturing" scientific products that make things "work."[17] For instance, she contends that scientific facts are constructed by a series of decisions made within such constraining influences as available research funds and facilities, idiosyncrasies in making laboratory preparations and applying laboratory procedures, social relationships with colleagues at the home research institution and in the special field, and the literary requirements of preparing formal research papers. Not all decisions and rejected alternatives are evident in a published research article. Instead, attention is focused on lab results treated as noncontroversial and irrefutable facts, as though what *has been done* in the research is all that *could be done*.[18] Here, rhetorical selectivity is at work in scientific activity.

The decisions scientists make during research take on a sharper rhetorical edge when scientists translate research activity through discourse addressed to other scientists. Kuhn implies that when a scientist makes discourse that reflects and guides perceptions according to the constraining influences of a chosen paradigm, a scientist is, by definition, engaging in rhetoric. Further, if we accept Knorr-Cetina's claims about the manufacture of scientific knowledge, rhetorical inducement is clearly evident in literary selections made to transform the results of laboratory work into seemingly noncontroversial and irrefutable facts. Hidden from the reader's view, however, are numerous choices and judgments; attention is focused on an image of research as a uniformly applied set of procedures that yielded singular results. Once more, persuasion is significantly involved. Scientific discourse reflects and guides perception in pursuit of what Perelman and Olbrechts-Tyteca called "adherence to a thesis."[19]

Henry W. Johnstone, Jr., offers a view of rhetoric that can sharpen our understanding of the rhetorical dimensions of paradigmatic scientific discourse. He contends that rhetoric is not concerned with influencing attitudes per se, but with developing strategies for "altering attitudes *in the service of propositions*."[20] Rhetoric is concerned with influencing attitudes in view of beliefs. Selectivity is, of course, intimately involved here. A rhetor's alteration of attitudes in the service of *some* propositions is simultaneously and necessarily alteration of attitudes at the expense of others. Given Johnstone's perspective, we see more clearly that paradigmatic discourse is selective rhetorical discourse in the service of a set of propositions that underlie the paradigm, and the discourse aims at having others adopt certain attitudes in view of those paradigmatic beliefs and values.

At its most fundamental level the discourse of a paradigmatic science is precisely an attempt to symbolize rhetorically so that one's favored paradigm is invoked and others are induced to cooperate with it in thought and conduct. This is why Kuhn is correct when he says that paradigmatic argument is not logically compelling for those who refuse to step within the "circle" invoked by the argument. The status of the circular argument "is only that of persuasion."[21] I would modify Kuhn's remark by dropping the word "only." There is no need so to minimize the integrity of what scientists do rhetorically. They do what they do because scientific method allows no alternative. Scientific discourse requires careful selection of symbols that will invoke a favored paradigm; colleagues must be induced to step within the paradigmatic "circle of understanding." Meaningful paradigmatic discourse cannot proceed otherwise.

Kuhn's description of doing science limits the scope of rhetorical persuasion to convincing another person that one's view is superior to alternatives.[22] However, if paradigmatic discourse is to be meaningful, a broader range of rhetorical inducements must be possible. At a minimum a scientific rhetor must induce another to think *at least tentatively* within a favored paradigmatic framework or there will not even be understanding. Intelligibility and understanding depend on the willingness of others to project themselves into that framework and on the capacity of that framework to depict problems and related phenomena coherently and meaningfully. Even if a respondent desires to refute a paradigmatic stance, he or she must accept tentatively the paradigmatic presuppositions in order to locate points for refutation.[23]

The logical circularity of paradigmatic discourse is but a special manifestation of a feature explained by Burke's general theory of symbolic orientation. Once we are induced to accept particular terministic screens, we gain entry to the orientation invoked by those terms. We will treat those terministic screens as unquestioned presuppositions, if only provisionally. We cannot enter the orientation otherwise. We may participate in the suggested orientation by following the attitudinal inclinations of terms expressed and by following out the conceptual, analytical, and evaluative implications contained in them. An orientation patterns or structures our experiences and prescribes how otherwise disconnected items should be related in sensible, orderly, and compelling ways. Of course, not all orientations are equally inviting. The strength of our commitment to an orientation will vary with the perceived capacity of its classifications and implied relationships to mediate otherwise ambiguous and chaotic features of experience and thereby provide a sense of what is to be expected in some experiential context and of what will fulfill those expectations.

All scientific discourse is rhetorical because it involves inducing others to share an orientation for evaluating and "making sense" of situated phenomena and the relationships among them. This is part of the rhetorical quest for meaning that is found in any problematic situation which invites resolving discourse, both within and outside the sciences. Burke's theory of symbolic orientation helps us understand that the kinds of orientations to which scientists are induced can vary in complexity and in flexibility of rhetorical applications. Orientations invoked through scientists' chosen terministic screens can thus reflect frameworks of thought from within scientific or other scholarly communities, or from the general culture. Orientations can be weak or strong in accounting for the contingencies of experience, with intermediate gradations. Orientations can provide the comfort of habit and convention or the intrigue of novelty and curiosity. They may supply noncontroversial options for engaging in scientific thought and conduct or they can compete or conflict with other "reasonable" alternatives that command varying degrees of adherence among community members. They may be general and tacitly accepted world views, fully and consciously articulated "master theories," or undeveloped heuristic analyses and working models with implications that have not yet been fully explored and worked out.

Orientations peculiar to special scientific communities of thought deserve to be called "paradigmatic" insofar as they reflect an evolved disciplinary matrix that sociologically constrains conduct of a given community's members. These ordered components of professional consensus induce communal identification among members of a specialty field and distinguish them from nonmembers.[24] Individually and collectively the constituents of a disciplinary matrix can be used as noncontroversial bases for invoking a paradigmatic orientation from which shared meanings and common understandings can be attached to otherwise potentially disputable or controversial claims and actions. Members of self-professing, professional, scientific disciplines cannot make communally meaningful discourse otherwise. This is so both within and outside of the natural sciences and regardless of comparable levels of scientific "maturation."[25]

But the special kind of paradigmatic orientation peculiar to a mature science includes within its disciplinary matrix concrete, subject-bound, problem-solving achievements that so constrain scientific work that they insist upon community consensus. This kind of paradigm defines a "normal science," governs the puzzle-solving activities of specialists in a mature scientific field, and does so by excluding alternatives as scientifically unreasonable. Alternatives can emerge in challenges to the paradigm's ability to yield concrete, subject-bound solutions for recurring and persistent problems—alternatives which, if successful, might generate new scientific specialties. The persuasive power of normal paradigmatic discourse comes in large part from its grounding in the professional habits of thought and action received through scientific training.

Scientific training is aimed at teaching students to think and to act as scientists, according to the orthodox beliefs, problem-solutions, and procedures of some substantive research area. A student must learn to perceive and to interpret research problems and related concepts and phenomena according to the prevailing paradigms of the field to which the student aspires professionally. Textbooks constitute the main communicative medium for propagating conventional ways of thinking about and doing a particular science, for inducing "a deep commitment to a particular way of viewing the world and of practicing science in it."[26] Textbooks present the accepted findings, terms, concepts, methods, and procedures of a field. This presentation typically provides a vision of a "normal" scientific paradigm as one fixed by nature rather than as one fixed upon nature by scientists' previous choices and deliberations.

Rhetorical inducement is involved in scientific education because the education offers a vision of a scientific specialty, usually without attention to possibly reasonable alternatives. This bias is not, of course, necessarily insidious. A student has a difficult enough task learning to think, to speak, to work, according to sanctioned paradigms, without having the additional burden of learning all about historical and unconventional alternatives. Nevertheless, scientific pedagogical discourse does induce students to think and make judgments according to the constraining influences of a paradigm and, to this extent, pedagogy selectively excludes alternatives. This kind of selectivity in science education is explained effectively by sociologist Barry Barnes:

> Scientific training always requires that a paradigm or paradigms be recognized as the sole legitimate representation of, and mode of dealing with, an aspect of the physical environment. It demands acceptance of the existing orthodoxy in a given field. Accordingly, it tends to avoid anything which might undermine or offer an alternative to that orthodoxy. The history of a field, wherein are found radically variant concepts, problems and methods of problem-solution, is either ignored, or is systematically rewritten as a kind of journey toward, and hence a legitimation of, present knowledge. . . . Similarly, current but unorthodox perspectives and procedures are overlooked; and possible weaknesses, or even well-known and generally recognized difficulties in orthodox interpretations, fail to find their way into teaching texts.[27]

A certified professional scientist trained in the prevailing paradigms of a field is ready to engage in research that extends or maintains the field's paradigmatic perspectives. This kind of activity in science is referred to (by Kuhn and others) as "normal science." Normal science is characterized by consensus that a certain paradigm is useful and valuable for solving problems of interest and for stimulating research possibilities. A scientist doing normal science may attempt to refine the paradigm by extending its scope and precision, but all normal scientific activity is undergirded by the assumption that a current paradigm supplies the means whereby puzzles may be brought to solution. As Kuhn put it, normal science involves attempts to fulfill a paradigm's promise of success:

> Normal science consists in the actualization of that promise, an actualization achieved by extending the knowledge of those facts that the paradigm displays as particularly revealing, by increasing the extent of the match between those facts and the paradigm's predictions, and by further articulation of the paradigm itself.[28]

Since shared paradigms constrain the range of acceptable and solvable problems, the consequences of normal research activity are largely anticipated at the outset, and sometimes in great detail. Scientists are challenged to apply their skills in empirical work designed to solve puzzles that are thought to be revealing from within a shared paradigmatic framework.[29] Normal scientific research strives to tighten connections between existing theory and existing observations, to make them "fit" more coherently and precisely. Normal scientific research sometimes tries to fulfill paradigm-induced expectations concerning the viability of extending existing theory to domains where it has not yet been applied. Normal scientific research is also concerned with collection of concrete data needed to extend and apply existing theory. Such "normal" tasks contribute to puzzle-solving activities in scientific research.[30]

Normal scientific discourse can be thought of as intraparadigmatic rhetoric. Scientists working intraparadigmatically are able to make and to evaluate rhetorical discourse, and they do so according to quite specific criteria: particular classes of phenomena are designated as suitable for experimental investigation, replicability, and measurability; terms and concepts are readily available for coherent depiction of experimental phenomena and for positing relationships among ideas; sophisticated problem-solutions are recognized as exemplars for modeling problem-solutions in novel research situations; particular laboratory procedures, methods, and preparations become standardized. There is, in brief, much about the shared paradigm that practitioners of normal science consider determinate and stable. The task of a normal scientific rhetor is therefore to induce assimilation of an indeterminate and potentially disputable claim into the comparatively determinate and stable paradigmatic framework. Of course, such claims must not merely be consistent with the embraced paradigm to gain acceptance; they must also be shown to extend or maintain that paradigm in accurate and fruitful ways.

All scientific research is not normal scientific research based on a shared paradigm. Leaving aside for the moment Kuhn's idea of paradigm replacement through intellectual revolution, I want to consider those research contexts and problems that pique the interest of scientists from different scientific specialties. Scientists will bring distinctive training and expertise to bear on such problems, and consequently paradigmatic perspectives are likely to make contact as discourse is carried out. How phenomena should be described and interpreted become issues of debate among investigators from different research specialties. Such clashes indicate a conflict of paradigms.

Interparadigmatic science occurs when research problems attract the attention of scientists who work from different paradigmatic perspectives. It is thus characterized by the absence of any single dominant paradigm commanding consensus among researchers attracted to the particular problem field. Three conditions will therefore be present if interparadigmatic science is to occur: (1) the absence of any single, shared, concrete, substantive, exemplary achievement for guiding investigations of the problem area; (2) pronounced difficulty securing consensus among researchers about which paradigm or combination of paradigms from among available candidates can further investigations; and (3) an apparent lack of "crisis" in any of the paradigmatic orientations proposed as means for solving the problems that generated interparadigmatic science in the first place.

The first two conditions distinguish interparadigmatic science from normal science. In interparadigmatic science, scientists make rhetoric about which orientations most meaningfully and usefully address research problems. Orientations proposed are likely to be weak, commanding much less collective adherence among scientists than those governing normal scientific activity. In the absence of shared, empirically based exemplars that apply directly and exclusively to the problem area, such orientations may be drawn upon to guide interpretations of particular research problems and, to this extent, can serve a paradigmatic function. The third condition distinguishes interparadigmatic science from revolutionary science. Unlike revolutionary paradigm change, acceptance of one orientation does not require rejection of another. Scientists from different disciplines may disagree about which orientations are most promising for tackling a research problem, but acceptance of one orientation does not spell doom for the others. Several orientations can coexist. What is at issue in interparadigmatic science is which paradigmatic orientation is most serviceable in solving problems of interest to two or more scientific disciplines. Each paradigmatic orientation still promises to guide investigation of problems within its own discipline, regardless of how well it fares in addressing interdisciplinary problems.

Rhetorical discourse is likely to take one of three forms in interparadigmatic science. First, scientists may address their research claims to targeted conferences, journals, or specific referees likely to embrace their preferred paradigmatic perspective. Scientists can try to make intraparadigmatic rhetoric by attempting to induce acceptance of claims as extensions of a paradigm shared with a targeted audience. In effect, scientists can choose to preach to those already saved. For instance, scientists can found new journals and organize special conferences to feature research pursued according to a particular paradigm, even though several other paradigmatic candidates exist for guiding solutions of the general problems of interest.

Second, scientific rhetoric may try to induce other interested scientists to accept a chosen paradigm as comparatively superior. They may argue for a particular paradigmatic orientation by appealing to values that scientists share *as* scientists, regardless of specialized discipline. For instance, they could urge acceptance of a chosen orientation because it supplies theoretical constructs, data, or investigative techniques that are, presumably, more fruitful, more accurate, or more precise than those of competitors.[31]

The third form of interparadigmatic discourse demands more explanation than the other two. This is a form of cooperative scientific discourse that takes place when scientists from different disciplines recognize they have interdependent interests in solving problems. Scientists from one discipline are seen to possess an orientation that can serviceably address the problems of scientists in another discipline, and vice versa. Perception of interdependence can lead interested scientists to deliberate about a new paradigmatic orientation that meets the needs of all those concerned.[32] The chromosome theory of Mendelian heredity offers a case in point. Researchers working in the fields of genetics and cytology (or cell biology) were interested in explaining the phenomenon of heredity. However, each group had questions not answerable with the techniques and concepts of its own field. Geneticists were interested in finding the location of genes. Cytologists, not geneticists, had the concepts and microscopic techniques needed for locating genetic material within the cell. Cytologists needed to investigate the functioning of chromosomes in producing individual hereditary characteristics. Geneticists, not cytologists, used techniques of artificial breeding and sought to explain patterns of inherited characteristics with postulated genes. This interdependence between cytologists and geneticists motivated the forging of an interfield theory that was serviceable in addressing the problems of both groups and opened up a new line of research coordinating the two fields.[33]

In sum, the rhetoric of interparadigmatic science can, but need not, culminate in normal science or in intellectual revolution. In principle the accuracy and fruitfulness of any paradigmatic orientation can vary with research contexts and related problems. Therefore this kind of scientific discourse can aim at producing fruitful coexistence, competition, or cooperation, rather than at coping with an authoritarian status quo on the one side or intellectual revolution on the other.

Accepting a revolutionary scientific paradigm is tantamount to rejecting the old or established. By adhering to a revolutionary paradigm one accepts that it at least promises that it at least promises that it can supply substantive, concrete, exemplary solutions to problems which led the old paradigm into crisis. Therefore the new ought to supplant the old. Conversely, by adhering to the old paradigm one must deny that problems confronting the established paradigm are significant or insist that the revolutionary alternative either cannot solve them more effectively than the old paradigm or cannot deal successfully with significant problems of its own.[34] The rhetoric of scientific revolution does not aim at cooperation, coexistence, or even friendly competition, then; acceptance of one of the alternative paradigms is tantamount to denying the legitimacy of the other's reasons for existence.

Revolutionary paradigmatic change can be elucidated by reference to Burke's theory of symbolic orientation. We all choose what to think and to say according to our overall conceptions of the world. When orientation-induced expectations are continually throttled in experience, people sense something has gone wrong. The foundation of reasonableness the old orientation once supplied seems to crumble. Those operating within the old orientation begin to feel, quite literally, disoriented. This mental state parallels that period of scientific change that Kuhn characterized as a "crisis" period, generated through the emergence of anomalies. Recognition and evaluation of anomalies depends on scientists' firm commitments to systematic scientific tradition. Those that are significantly disorienting are thought to be in "explicit and unequivocal conflict with some structurally central tenet of current scientific belief."[35] Revolution-inducing anomalies violate fundamental paradigm-induced expectations persistently; they "penetrate existing knowledge to the core."[36] Only this sort of anomaly can shake scientists' confidence in a scientific tradition and evoke contemplation of paradigm change.[37] When this happens, a crisis exists in science—a dramatic instance of what Burke's more general theory defines as disorientation.

Kuhn has pointed to contemporary examples of disorientation generated by a paradigm crisis. Einstein wrote: "It was as if the ground had been pulled out from under one, with no firm foundation to be seen anywhere, upon which one could have built." Before Hiesenberg's paper on matrix mechanics suggested a course toward a new quantum theory, Wolfgang Pauli wrote: "At the moment physics is again terribly confused. In any case, it is too difficult for me, and I wish I had been a movie comedian or something of the sort and had never heard of physics."[38] The disorientation reflected in such "crisis" statements is plain.

Generally the demise of an orientation is signaled by its inability to supply meaning and by the emergence of disorientation. A shift of perspective, a reorientation, may then be imminent. Terms that invoke the old orientation are challenged, and rhetorical invitations to cooperate and participate with the old orientation are increasingly denied. Instead, individuals seek out new terms through which to render the ambiguous experiences meaningful, terms that hold promise for reorienting participants. The persuasiveness of any new orientation lies in the extent to which it promises a new sense of relationship by rendering what was puzzling or inconsonant meaningful and consonant. These processes do not occur only in science; that they occur generally indicates that Kuhn's theory of paradigmatic change in science is not esoteric or field specific.

Revolutionary paradigm change can be understood as a specific, subject-bound occurrence of symbolic reorientation. Some of a scientific field's fundamental goals, problems, theoretical propositions, methods, and standard applications will be altered significantly during the process of revolutionary paradigm reconstruction.[39] Examples of such revolutionary paradigmatic changes that Kuhn cites include the triumph of the Copernican system over the Ptolemaic system in astronomy; the transition from Aristotelian to classical mechanics and the movement from traditional mechanics to quantum mechanics and relativistic mechanics in physics; and the replacement of the phlogiston theory with the oxygen theory of combustion.[40]

A point to stress is that during periods of scientific crisis, rhetorical discourse will display some standard strategies of persuasive argumentation that are not prominent in normal science or in interparadigmatic rhetoric about orientations. Advocates of a revolutionary paradigm will argue that the current paradigm is in a state of crisis due to the presence of anomalies fatal to the paradigm's core presuppositions. Further, they will strive to gain at least tentative acceptance of a revolutionary paradigm by offering it as a *progressive* alternative. To accomplish this aim, revolutionary rhetors might fashion rhetorical appeals that draw upon, transform, and uniquely extend applications of traditional beliefs, opinions, values, and general notions.[41] This could involve developing fertile metaphors and analogies that link notions shared by controversialists as members of a general culture with more controversial statements about technical problems or possible solutions that concern them as scientists. Sometimes metaphors and analogies might be used to connect notions regarding an established principle in one field of endeavor (scientific or otherwise) with a disputed principle possessing revolutionary implications in another. For instance, Darwin sought to advance the principle of natural selection through an analogy based on the more generally known and acceptable principle of artificial selection of traits in domestic breeding.[42] In response, defenders of the status quo are apt to argue that the anomalies are not significant enough to present a crisis. Or they may contend that the cited anomalies will, with time, be accommodated by the current paradigm. Or they will sometimes argue that more will be lost than gained by replacing the old paradigm with a comparatively underdeveloped alternative. Further, they might challenge efforts at creatively transforming commonplace understandings of received beliefs, opinions, and values. They will deny the appropriateness of metaphors and analogies that proponents of the revolutionary paradigm frame. These lines of argument are prominent in rhetoric about revolutionary change because revolutionary discourse, scientific or otherwise, aims firmly at acceptance of one paradigm and final rejection of its competitors. I turn next to the rhetorical dimensions of scientific discovery.

According to Kuhn, scientific discoveries fall primarily into two classes: discoveries predicted in advance from accepted theory and discoveries unanticipated by accepted theory.[43] Discoveries of the first class are made with considerable understanding of what to look for. Examples include the discoveries of the neutrino, radio waves, and elements which filled empty places in the periodic table. Discoveries of the second class violate the expectations induced by a standard paradigm. Their acceptance requires more than additive adjustment of existing theory; assimilation of those novelties requires revaluing, reordering, and transforming an old paradigm in ways that could contribute to or cause what Kuhn characterized as intellectual revolution. Discoveries in this class include oxygen, Uranus, and x-rays.

How did such occurrences come to be called "discoveries"? Augustine Brannigan's attributional model of scientific discovery supplies an answer (and it is rich in rhetorical implications). According to Brannigan, naturalistic approaches that search for psychological or historical causes do not offer completely adequate explanations of scientific discovery. Scientific discovery is better investigated as a category of social approbation. Brannigan therefore delineates social features that must be present in a situation where claims to scientific discovery are intelligible to other scientists. His model may be summarized thus:

> Rather than treating discovery as a naturalistic occurrence which requires a naturalistic explanation, in this account we have examined the features of intelligibility of the phenomenon of discovery which appear to ground the perception or constitution of discovery. The attribution of the status, discovery, is founded on the processes of social recognition by which the announcement of an achievement is seen to be a substantively

> relevant possibility, determined in the course of motivated scientific investigation or schemes of research, whose conclusion or outcome is convincingly true or valid, and whose announcement is, for all appearances, unprecedented. These are the central elements in the apprehension of scientific discoveries, both for the individual scientist and his or her community.[44]

Rhetorical communication is thus intimately involved in the assessment of discoveries. An audience of scientists must be induced to attribute the qualities of "discovery" to a claim. A scientific rhetor making a claim to discovery will attempt to constrain the audience's judgments through a series of selected items of information and argument that together make up an announcement—a major claim. Brannigan indicates the importance of rhetoric in establishing scientific discoveries as "discoveries" when he observes that only a selective class of details from among a broad and indefinite array of experiences can be used while making discovery announcements.[45] A scientist trying to secure a discovery claim must make rhetorical selections which promote perception of the discovery claim as fitting criteria of attribution: Such a claim must be *possible* or *desirable,* given the conventional state of a field's knowledge; must be perceived as a result of *scientific* work; must be *veracious;* and must have the quality of *uniqueness*. Brannigan points out that these same criteria constitute

> the elements by which claims to discovery will be ignored or disputed; consequently, claims that an announcement is *not* "news" or is hardly likely to be true, or is patently incorrect, will operate as a set of invalidation procedures for the disqualification of candidate achievements.[46]

Thus scientific discourse is rhetorical in that it seeks to induce in one or more of three ways. Terminological selectivity necessarily argues for the legitimacy of a particular symbolic representation. This level of persuasiveness is inevitable in any resort to language, scientific or other. The discourse will also invite at least tentative cooperation in pursuing a subject or problem from the general orientation implied by the terministic screens used in the discourse. Discourse will draw attention to a particular kind of description or vantage point and deflect attention away from (though not necessarily deny) alternative ways of looking. Finally, the discourse will imply or even demand acceptance of a particular paradigmatic way of inquiring, proceeding, and judging. When shared, concrete, substantive, exemplary problem-solutions govern an orientation underlying discourse, that orientation is the most insistent and exclusory. It must be added, however, that successful inducement at any of these levels is bound to be constrained by situational variables.

SCIENTIFIC DISCOURSE IS SITUATIONAL DISCOURSE

To this point I have argued that a paradigm predetermines one's orientation toward what allegedly is and can be. Now I contend that a paradigm defines what can and cannot be said *situationally*. For example, within an S-R or even an S-O-R paradigm of human behavior, concepts of soul or mind have no place, regardless of where discourse occurs; but within a psychological paradigm based on cognitive processing, mental processes not purely reflective of electrical circuitry may be posited, regardless of situation. In such ways scientific paradigms define exigences and constraints of what in rhetorical theory is called "a rhetorical situation."

Further, any paradigm that constructs a scientist's conceptions of how to think and how to do science stands in a historical context. Scientific claims are proferred and evaluated within the context of what scientists know and value at a particular time. This is so regardless of whether recommended changes in what is known and valued are incremental or revolutionary. Scientific knowledge advances through discussion and debate over the best way to think and to do science, and it does so within particular settings that constitute rhetorical situations. Any instance of symbolic reorientation in science requires suasory processes that are continually adjusted to specific audiences addressed in particular times and circumstances. Even the components of a consensus that strengthen adherence to a normal scientific paradigm were forged at some time through debates, discussions, and adjudicating judgments constrained by situations.

Logic and experimentation are not the fundamental means of securing scientific change. They are efficacious only if applied *persuasively*. Scientific advancement is the result of dynamic and interdependent relationships between scientists' efforts at persuasion and adjudication by audiences in actual and specific, temporal and physical situations. Therefore, we need to explore the processes by which this discursive activity is carried on.

Like every rhetor, a skilled scientific rhetor must create discourse that addresses or brings into view an exigence, or ambiguity, whose modification is important to the specific audience addressed. Further, a scientific rhetor must take into account the opportunities and limitations (constraints) of the specific situation in which discourse will occur. The scientific rhetor must locate those theories, opinions, and values accepted by the prospective scientific audience; where such contents can be had, they will minimize misunderstanding and disagreement. In the language of rhetorical theory and tradition, a proposer of change should seek the largest possible "common ground" in order to render the points of clash, the *stases,* as few and narrow as possible.

Research settings constrain what can be said and what can be offered in support of claims. Discourse is constrained by such variables as the availability of financial resources, the quality of available research facilities and equipment, or the technical competencies and interests of research colleagues. For example, if a scientist has minimal funding or minimal research equipment, this limits what he or she can investigate and propose. In scientific communities it is common to criticize "hammer and nail" research in which, having a particular set of tools, the scientist chooses to investigate what can be investigated with those tools rather than those things that most need investigation. In practice, however, all scientific discourse derives from what scientists actually have to work with. Originality is not exclusively the product of intellection; it is also a product of available means of investigation.

The character of scientific discourse is further affected by the reputation of the researchers in their specialties. A scientist not already recognized as competent within his or her community needs to work rhetorically to establish credentials. An established scientist needs to expend little rhetorical effort in this direction. It is also the case that when a scientist addresses a technical exigence that is not already widely recognized by the community addressed, that scientist must make rhetorical efforts to establish the importance of that problem in order to gain attention for the scientific claims offered.

Such variables are more extensive and detailed than those indicated. Indeed, there is a large body of sociological literature about such constraining influences on scientific activity, especially the influence of funding and the politics of research support.[47] The existence of such constraints on scientific rhetoric shows clearly that rhetorical success in scientific activities is subject to the influence of situational variables. At the minimum, scientists' public discourse must show command of the requisite resources and competencies for conducting research that promises to modify the scientific exigences addressed.

Scientists are subject to rhetorical constraints whenever they select technical exigences for investigation. Exigences selected for study must be exigences that the scientific community addressed sees as legitimate, or the exigences themselves must be argued for.[48] Technical exigences always become *rhetorical* exigences when chosen with a view to evoking satisfactory responses from other scientists. A scientist will not choose just any available problem but will choose one that, given personal competencies and available resources, he or she is likely to modify and to receive credit for modifying.[49]

Selection of technical exigences is further constrained by a specialty's state of conventional knowledge. As I have already said, all scientific advancement—incremental or revolutionary—is both product of, and response to, developing scientific tradition. The conventions of a scientific specialty constrain what can be successfully inserted into the developing tradition of that science.

Latour and Woolgar argue that the process of constructing scientific facts employs devices that disguise the reality that the "facts" have been produced by scientists rather than given by nature.[50] These authors contend that even laboratory equipment used to conduct experiments is the materialization of decisions made at some earlier time.[51] Such observations emphasize the stable and determinate posture scientific conventions lend to the presentation of "facts," a stable posture possible only because each such fact is a culmination of suasory inducements and judgments that occurred at specific places during particular periods of time. Thus, a scientific specialty's conventional knowledge constrains precisely because the history of argument and counterargument that gave conventional knowledge birth has now ceased to be a focus of attention. Any claims specialists put forward must be consonant with unstated agreements which constitute the conventional knowledge of their specialty.

It is imperative that each scientific rhetor decide which components of any conventional paradigm can safely be treated as stable and which must still be treated as subject to change or development. Without such distinctions technical claims cannot be reasonably inserted into the ongoing deliberations of a field. It is ironical that the conventions of scientific reporting encourage an authorial stance that promotes an image of technical data and claims as stable and determinate. Data and claims are not usually presented in ways that call attention to the contexts in which decisions and judgments gave rise to them. Instead, readers are invited to conclude tacitly that what the research scientist has said and done was all that *could* be said and done given the research situation and the (presumably) fixed state of related knowledge.[52]

Scientific articles are often subtly persuasive because the manner of presentation diminishes readers' awareness of the selectivity that operated during research activity and in the formation of the paradigmatic tradition within which rhetors and audiences stand. Articles typically describe materials, equipment, and procedures; they refer to literature alleged to contain certified scientific facts and viable hypotheses; they apply labels, taxonomies, and definitions. Yet, as presented, this content tends to be divorced from the actual research contexts that gave rise to it. Reports of false leads and unsuccessful procedures are frequently omitted from the text. So is discussion of viable alternatives to the decisions the reporting scientist made. By these rhetorical processes, selected choices are highlighted and alternatives are muted or omitted. Research articles are thus more persuasive precisely because the articles present an image of determinacy in science and obscure the influences of historical and situational circumstances and the possibilities of alternative approaches and judgments.

Formal scientific writing conventions frequently have the further rhetorical effect of implicitly proclaiming the fixity of conventional knowledge and the objectivity of inquiry. Scientific research papers tend to be written in an inductive style that implies that claims were found through impartial investigation of phenomena that have independent, objective, and undeniable existence. Findings are "reported" more than argued for. Certain favored stylistic conventions also contribute an aura of unsullied inductive procedure and objectivity. For example, passive-voice constructions are common and in some cases are demanded by style sheets; hence, we read "It was observed that . . ." rather than "I (we) observed that. . . ."[53] Halloran refers to the strategy of manufacturing "abstract rhetors." Authors do not show or argue "the *data* show that . . " or "This *paper* will argue that. . . ." The cumulative effect of such rhetorical strategies is assertion or implication that what is said and done is all that *could* be said or done within the research situation. Some scientific communities have begun to discourage the most transparent of these stylistic strategies:

> Absolute insistence on the third person and the passive voice has been a strong tradition in scientific writing. Authorities on style and readability have clearly shown that this practice results in the deadliness and pomposity they call "scientificese." Some scientists maintain that this style preserves

> objectivity, but the validity of this assertion is suspect. . . . An experienced writer can use the first person and the active voice without dominating the communication and without sacrificing the objectivity of the research. If any discipline should appreciate the value of personal communication, it should be psychology.[54]

My concern here is not so much with what styles scientists choose in their reporting as with the undeniable fact that scientific reporting is *rhetorical* in the sense that styles are chosen because of their presumed persuasiveness concerning the virtuosity of science itself.

The fact that one or another style is preferred for scientific discourse is clear indication that scientific discourse is adapted discourse—adapted both to persuasive purposes and to the expectations of specific audiences.[55] I do not wish to imply that scientists are slaves to stylistic convention. Rather, style in scientific discourse always has rhetorical purposes and functions, and these are often minimally connected with either the content of what is said or the actual circumstances under which an investigation proceeded.

The most thoughtful scientists know that without selectivity there could not be innovation in science, whether innovation involves puzzle solving or intellectual revolution. The important point is that selections of procedures, data, or style are always in *response* to something or someone. They are adaptations to real or imagined exigences, and insofar as the adaptations are not accidents they involve thinking rhetorically.

A scientific rhetor tries to insert technical claims into the developing literature of a field by noting or creating *stases* and recommending solutions that transform these indeterminate points into stable, determinate scientific knowledge. Technical claims need to be couched in larger bodies of content that the audience is likely to accept as indisputable. If an argument for a claim is to work, the bases of common ground must hold firmly for at least the time being. Knorr-Cetina stops just short of identifying the state of conventional scientific knowledge as the reservoir both for locating points of conflict *(stases)* and for finding commonplaces from which to argue in behalf of a position:

> If scientific objects are selectively carved from reality, they can be deconstructed by challenging the selections they incorporate. If scientific facts are fabricated in the sense that they are derived from decisions, they can be defabricated by imposing alternative decisions. In scientific enquiry, the selectivity of the

> selections incorporated into previous scientific work is itself a *topic* for further scientific investigation. At the same time, the selections of previous work constitute a *resource* which enables scientific enquiry to proceed: they supply the tools, methods, and interpretations upon which a scientist may draw in the process of her own research.[56]

The complex of previous decisions comprising conventional scientific thought and practice constrains scientific discourse. To advance claims successfully, scientific rhetors must be able to discriminate what the audience, with its background of knowledge of previous scientific decisions, is willing to accept as points of legitimate *stasis*. It is at these points scientific rhetors appropriately urge changes in thought and practice. Having located points of scientific *stasis,* the rhetor must next discover sources of common ground—*scientific* commonplaces—that can be used to propose change and make it seem reasonable within the particular rhetorical situation.

How an audience perceives the professional character of a scientific rhetor or group of rhetors is a further situational constraint. Aristotle's contention that a central means of persuasion is the rhetor's perceived ethos clearly applies in science.[57] The groundbreaking work on scientific ethos was done by Robert K. Merton.

Merton sought to identify the binding institutional norms which constrain the behavior of scientists and facilitate establishment and extension of certified objective knowledge of the physical world.[58] The norms of science, Merton asserts, are intended to minimize distortion during systematic observation and maximize efficient dissemination of certified knowledge. Merton describes such hoped-for ethos of science this way:

> The ethos of science is that affectively toned complex of values and norms which is held to be binding on the man of science. The norms are expressed in the form of prescriptions, proscriptions, preferences, and permissions. They are legitimized in terms of institutional values. These imperatives, transmitted by precept and example and reenforced by sanctions are in varying degrees internalized by the scientist, thus fashioning his scientific conscience or, if one prefers the latter-day phrase, his superego. Although the ethos of science has not been codified, it can be inferred from the moral consensus of scientists as expressed in use and wont, in countless writings on the scientific spirit and in moral indignation directed toward contraventions of the ethos.[59]

The norms identified in Merton's original formulation are as follows. *Universalism* requires that knowledge claims be subjected to preestablished, impersonal criteria that render them consonant with observation and previously established knowledge.[60] *Communality* prescribes that research is not a personal possession but must be made available to the community of scientists.[61] *Disinterestedness* requires that scientists strive to achieve their self-interests only through satisfaction in work done and prestige accrued through serving the interest of a scientific community.[62] *Organized skepticism* mandates that scientists take nothing on faith, temporarily suspending judgments in order to scrutinize beliefs critically against empirical and logical criteria of judgment.[63]

Merton added to his original list the norms of *originality* and *humility* when he studied priority claims in scientific discovery.[64] Originality is a source of esteem for one's work because through this quality scientific knowledge advances. Humility ensures that scientists will not misbehave at the rate they would if importance were assigned only to originality and to establishing priority in scientific discovery.

Much research done in the wake of Merton's studies has supported his functional norms,[65] but there also has emerged evidence that there exists in science a set of counternorms wholly incompatible with those norms identified initially by Merton. In a well-documented study of NASA "moon scientists" Mitroff adduced evidence for some of the counternorms.[66] He found that univeralism is countered by *particularism.* Scientists often regard it as legitimate to judge research reports on personal criteria, such as the ability and experience of the author, rather than on the technical merits of knowledge claims themselves. Similarly, communality is opposed by *solitariness.* Scientists do often exercise property rights over their work. When they do, secrecy can seem appropriate and, indeed, can even help scientists avoid disruptive priority disputes and ensure that a scientist does not waste colleagues' time by rushing immature work into print. Further, disinterestedness is countered by *interestedness.* This counternorm prescribes that scientists try to achieve their self-interests through serving special communities of interest. Finally, organized skepticism is balanced by *organized dogmatism.* From this view, scientists do not doubt their own and others' findings incessantly, but instead assent fervently to their own findngs while doubting the findings of others. To put the point more positively, many scientists believe that to be a scientist is to exercise expert judgments given incomplete evidence.[67]

Evidence of conflicting commitments among scientists has led some sociologists to question whether the constituents of Merton's scientific ethos should be considered normative at all.[68] This question remains unsettled, but I do not intend to address it here.[69] I do contend that scientific ethos is not given; it is constructed rhetorically. Rhetors respond to, or seek to avoid creating, ambiguities and conflicts about their scientific credibility. They do this by choosing strategic options best suited to situational contingencies. What sociologists of science have been calling the norms and counternorms of science are effectively conceived as rhetorical *topoi* that index the available range of discursive strategies for establishing negative or positive audience perceptions of scientists' ethos.

The constituents of scientific ethos become salient only when the discourse of one scientist is evaluated by others in scientific situations that are rhetorical—that is, in problematic or ambiguous situations that involve inducing adherence to ideas as "scientific." Within such situations a famous passage from Aristotle's *Rhetoric* applies:

> Persuasion is achieved by the speaker's personal character when the speech is so spoken as to make us think him credible. . . . It is not true . . . that the personal goodness revealed by the speaker contributes nothing to his power of persuasion; on the contrary, his character may almost be called the most effective means of persuasion that he possesses.[70]

Acceptance of scientific ideas hinges on the perception that the proposing thinker or researcher possesses qualities of admirable scientific character, and what qualities will be most valued can vary from rhetorical situation to rhetorical situation.

Secrecy, communality, objectivity, and emotional commitment may or may not be situationally relevant to evaluating a scientist's professional work. The situation in which scientific substance is being evaluated determines the relevance and salience of each such quality. When a scientist makes knowledge claims to colleagues and implies that these claims are "scientific," it *may* be relevant, though it will not always be imperative, to weigh the claimant's objectivity or emotional involvement, openness or secrecy, skepticism or enthusiasm, and the like. In various situations either of such opposed qualities may be relevant and either may be judged a scientific virtue or vice.

What is said about a scientist's conduct and what is displayed about him or her in reportage may well differ—generally and from situation to situation. Points of importance are: (1) Reputation and perceivable conduct will be present wherever scientists communicate with each other;[71] (2) if there is any ambiguity or cause for doubt about claims, factors of ethos will be weighed in formation of final judgments; and (3) character per se will be limned in some degree in all reportage. For example, if a scientist says he or she double-checked a finding, that remark tends to enhance the qualities of thoroughness and skepticism as components of the scientist's ethos.

As soon as we recognize we are dealing here with rhetorical *topoi*, the seeming conflict between scientists' norms and counternorms dissolves. The themes having to do with scientific ethos are not lawlike or formal rules; they are alternative lines of thought that can bear on a scientist's credibility in this or that rhetorical situation. Mulkay's description of how features of scientific ethos come into play reflects their rhetorical topicality:

> In science . . . we have a complex moral language which appears to focus on certain recurrent themes or issues; for instance, on procedures of communication, the place of rationality, the importance of impartiality, and of commitment, and so on. But . . . no particular solutions to the problems raised by these issues for participants are firmly institutionalized. Instead, the standardized verbal formulations to be found in the scientific community provide a repertoire or vocabulary which scientists can use flexibly to categorize professional actions differently in various social contexts.[72]

"Standardized verbal formulations" translates in traditional rhetorical language into "*topoi* for rhetorical treatment." The formulations are repeatable and acceptable themes that deal with shared beliefs, values, and opinions. These "recurrent themes or issues" have to do with situationally appropriate scientific thoughts and actions. However briefly or amply treated, they refer to *proprieties* of doing science.

Which *topoi* seem most relevant for amplifying qualities of a rhetor's thought and conduct as *scientific* will be influenced by prevailing situational constraints and contingencies. Any of the *topoi* can be used to build or diminish perceptions of a rhetor's scientific ethos if situationally relevant, but there is nonetheless a rhetorical logic at work that favors "Mertonian" *topoi* over the "counter" *topoi* in at least some recurring kinds of rhetorical situations. When defenders of scientific orthodoxy directly and publicly challenge the scientific ethos of radical or unconventional claim-makers, *topoi* like universality, communality, skepticism, and disinterestedness are likely to figure prominently in argumentative attacks. These and related *topoi* characterize conventional ways of thinking about scientific qualities of thought and conduct. The logical principle of *presumption,* therefore, will usually rest with these qualities as constitutive of scientific ethos. Defenders of scientific orthodoxy can thus use them as powerful persuasive means of discrediting radical claims and claim-makers. This does not mean that radical claim-makers cannot turn to alternative themes, such as particularity, to induce more favorable perceptions of their scientific ethos; however, there is greater likelihood that they would have to meet a *burden of proof* in establishing such counter *topoi* as situationally relevant and constitutive of scientific ethos. In any case, *topoi* are chosen in accordance with a rhetorical logic for locating the grounds of assessing what constitutes situationally relevant and scientifically reasonable qualities of thought and conduct.

I have not exhausted the ways in which scientific discourse is constrained by features of rhetorical situations. Adoption of a paradigm predetermines what exigences deserve to be met in given contexts. In any context any scientist will be constrained by what prospective audiences can be presumed to know and value. Research settings influence what technical exigences can be usefully addressed, hence what kind of discourse can ultimately be inserted into the overall flow of scientific discussion. The state of conventional knowledge in any scientific specialty directs what can be addressed and how, at this time. Varying conventions of scientific reporting constrain those who report. Conventional and situational features constituting what will be perceived as praiseworthy scientific ethos will vary, making some *topoi* associated with ethos salient in some situations but not in others. In all of these ways scientific activity and especially scientific discourse are subject to limitations and opportunities that are situational, and therefore essentially rhetorical. Implicit in this review of rhetorical dimensions of scientific activity is the further fact that scientific discourse is always *addressed* discourse.

SCIENTIFIC DISCOURSE IS ADDRESSED DISCOURSE

Whenever scientific activity culminates in discourse, the message will be addressed to an audience, and that fact inextricably involves symbolic inducement that is amenable to rhetorical analysis.[73] I do not contend that all doing of science is rhetorical, but any activity that requires an encounter with other minds becomes in some degree rhetorical. Some have argued that even an individual's private reflections constitute internal discourse addressed to the self with the aim of self-persuasion. This casts too wide a conceptual net and denies us some useful, clarifying distinctions. Suppose that through private reflection a scientist becomes convinced that a particular claim has sufficient experimental support to justify its being scientific knowledge. In my view this is not of rhetorical interest because it does not involve addressing other minds.

The distinction I am making is important because within contexts of scientific discovery and innovation some scientists become convinced their views are correct before they address others. Albert Einstein provides an excellent example of a scientist thinking in nonrhetorical ways without concern about audiences. In 1919 he received a telegram that contained the results of an expedition that had gone to Principe to test Einstein's relativity theory by measurements of a solar eclipse. A student companion was delighted with the good news but Einstein was not, saying that the theory was correct all along. Indeed, when asked how he would have responded if the results had not confirmed his theory he replied, "Then I should have been sorry for the good Lord, for the theory is correct."[74] The story illustrates a mind at work without concern for rhetorical considerations of audiences.

If we restrict "the rhetorical" to that which is addressed to other minds, the range of rhetorical discourse in science is still vast. To do science as actively participating members of a self-defined scientific community, scientists *must* become rhetors. One obvious reason: Professional rewards and achievements are tied to gaining public appraisal and positive authorization for work completed. Another: The responses of others are helpful in developing and refining technical claims. Simply put, science as an enterprise exists by virtue of *shared* knowledge. Scientific knowledge in any field grows through dissemination and evaluation within the scientific community, and both processes involve rhetorical address to other minds. Despite Einstein's example, most scientists cannot gain assurance about ideas and methods without giving them voice through discourse addressed to others. It is by addressing others and interpreting their responses that most individuals find assurance they are perceiving and interpreting experiences in ways that make sense.

Generally speaking, the rhetorical audiences of science are members of scientific communities. These audiences are variable. Taken in large, we can consider the general body of scientists as a rhetorical audience to whom scientific communicators must adapt if they are to be accepted as doing science. When functioning as a scientist, a rhetor must always be conscious of what is required to show that he or she can think and act scientifically. A scientific rhetor addressing a broadly based professional conference may also have to make specific adaptations to, say, molecular biologists or nuclear physicists included among the auditors. However, the most significant rhetorical audiences for most research scientists will be those groups of people with whom he or she must interact regularly because of shared technical interests. Members of such an audience are likely to belong to the scientists's "invisible college" because they share similar research interests, problems, professional networking, and standards.[75]

In their professional work scientists think strategically about what claims will be acceptable to the scientific audiences they value most. Scientists share with professional academics and other scholars high critical standards for assessing arguments relevant to their respective disciplines. This means that most scientists think rhetorically about audiences throughout the development of their research work. Problems are chosen according to whether they are likely to be judged significant by the most valued audience. Procedures and preparations used in the laboratory are judged for their professional acceptability. Corroborative facts are drawn from published sources likely to induce support by their established credibility. Whether a scientist is working in the laboratory, reflecting on problems in the study, or preparing a final draft for publication, considerations of audience will permeate his or her thinking.

Indeed, the range of decisions made in adapting to an audience goes beyond the singular need to adjust to the audience's critical standards, however important that may be. Knorr-Cetina describes some of the additional audience adaptations a scientific rhetor must make:

> If we look at the process of knowledge production in sufficient detail, it turns out that scientists constantly relate their decisions and selections to the expected response of specific members of this community of "validators," or to the dictates of the journal in which they wish to publish. Decisions are based on what is "hot" and what is "out," on what one "can" or "cannot" do, on whom they will come up against and with whom they will have to associate by making a specific point. In short, the discoveries of the laboratory are made, as part and parcel of their substance, *with a view toward* potential criticism or acceptance (as well as with respect to potential allies and enemies!).[76]

A central rhetorical function of the scientific audience is to serve as an authorizing agency in respect to claims addressed to problems of concern to the community it represents. Authorization can come about without total commitment to claims made. Scientific audiences, as "gatekeepers," can grant the reasonableness of claims—can *appreciate* an investigator's claims—without believing the claims constitute final, scientific truth. Scientific authorization of claims can derive from agreements that a rhetor's claims are plausible enough to deserve further consideration within the community. The reasonableness of scientific rhetoric derives from audience judgments about the scientists' mastery of relevant data and skill in argumentation. A fresh claim can, of course, be accepted in full; but the "career" of a scientific claim could also be first appreciative understanding, then further discussion, refining and testing by claimant and authorizers, and finally confirmation of the original claim and incorporation into the official body of scientific knowledge.

Rhetoric that seeks authorization of a scientific claim must be addressed to an audience of scientific specialists who deal regularly with the subject matter of the claim and who can legitimate that claim as knowledge. Members of each scientific community recognize their community as the "exclusive arbiter of professional achievement."[77] Appeals to political authorities or to the populace at large, for example, are taboo in doing science. So too are appeals to scientists whose technical competencies are suspect. (This would be the case with scientists working in specialties markedly different from the claimant's specialty.) Students constitute another audience judged unqualified to evaluate scientific claims with a view to their authorization or rejection.[78] In sum, to secure appreciative understanding of knowledge claims and to secure resolution of rhetorical exigences in science, a scientist must submit work for authorization by those who represent "the well-defined community of the scientist's professional compeers."[79]

Rhetorical adaptations to the expectations and standards of a scientific field are relatively predictable. The membership of any given scientific speciality is homogeneous. Common interest in solving particular kinds of problems is one factor producing homogeneity. Professional training is another. Like training inculcates like values, beliefs, and experiences. Scientific rhetors know this, and make rhetorical adaptations accordingly.

In science authorization of new claims requires that a rhetor tacitly or explicitly affirm commonplace concerns held by the scientific audience addressed. Claims must be grounded in that which the community considers reasonable. Scientific discourse is rhetorical discourse because, among its other rhetorical features, it must offer content situationally defined as reasonable.

SCIENTIFIC DISCOURSE IS REASONABLE DISCOURSE

As I have indicated several times, I agree with Kuhn that scientific claim-making and evaluation do not proceed on the basis of formally logical analysis alone.[80] For one thing, arguments that possess formal validity are not always persuasive, even with a qualified audience. One can disagree with arguments for reasons other than those that arise from stipulating and applying formally logical criteria. For instance, we may challenge the antecedent premises of a logical argument or challenge the values those premises imply. Formally logical criteria are insufficient for making and evaluating scientific claims, but this does not mean that all efforts to secure or resist scientific change are left bereft of standards and are thus irrational or unreasonable, as some have claimed.[81]

Kuhn's discussion of conversion and related psychological accounts of scientific change has perhaps obscured a clearer understanding of scientific discourse as discourse judged reasonable or unreasonable *in rhetorical situations.*[82] People do use symbols to induce cooperative acts and attitudes, and to do so is to make rhetoric. A rhetor, whether scientist or other, attempts to induce the participation of an audience in his or her favored orientation and related claims. The results of successful rhetoric can entail varying degrees of adherence, so rather than use Kuhn's concepts of neural reprogramming and conversion, I suggest that when scientific discourse occurs, those who respond to it *choose* the degree to which they will adhere to propositions offered. In a scientific community choices will be made with due respect for the values and knowledge problems shared by that community.

Whether it takes place among proponents of different paradigms or adherents to the same paradigm, scientific debate and discussion tend to be about premises, the relevance of data, and the comparative worth or particular applications of "good reasons" put forward to gain acceptance or rejection of claims. Criteria for accepting premises, for determinig relevance, and for deciding which reasons are "good" are found in communally shared standards for governing scientifically reasonable claim-making. Inasmuch as these criteria reflect the problem-solving expectations and valuational considerations of scientific communities, they are much broader than the rules of formal logic and, indeed, encompass them. However, when any such criteria are invoked—including criteria of formal validity—they do not function like neutral algorithms for determining choice, but rather as values for situationally constraining and influencing choice.[83] A scientific community can share many values and still argue legitimately about their comparative importance and applications to particular cases. This is where scientists must have recourse "to persuasion as a prelude to the possibility of proof,"[84] and where they do, the logicality of persuasive success depends on whether those claims can be judged situationally reasonable.

One of the most important questions members of scientific communities raise when something is offered as science is: Will that help us solve any of *our* significant knowledge problems? If the message seems potentially helpful, the discourse in which the propositions were embedded will be thought reasonable as science. Thus, scientific discourse is evaluated against experimental and logical criteria but also, and perhaps most tellingly, according to its apparent promise in fulfilling the dominant motive for doing science—solving problems.

But can scientists operating from different paradigms take the measure of each others' potentialities for solving relevant and significant problems? With Hilary Putnam, I believe they can. Putnam challenges the incommensurability thesis, articulated in the writings of Kuhn and, more radically, of Feyerabend.[85] Among his claims is the contention that if there were true incommensurability among cultures, we could not translate the languages of other cultures in ways that "make sense" of their beliefs and knowledge. Yet we do this all the time. We make sense of seventeenth-century scientists even though they belonged to another culture. Putnam concludes that the incommensurability thesis is ultimately self-refuting, using among his examples Feyerabend's application of the idea to Galileo's science. Said Putnam: "To tell us that Galileo had 'incommensurable' notions *and then go on to describe them at length* it totally incoherent."[86] Putnam's critique implies there is potential for making genuinely reasonable interparadigmatic discourse in science. That competing proponents of paradigms choose and work from different conceptions of how to think and to do science does not mean they cannot learn, understand, and assess the reasonableness of each other's paradigms. Interpretive success during interparadigmatic discourse does not require establishing meanings and conceptions that remain the same as we move from one paradigm to another. The ability to understand even a comparatively bizarre culture's conceptions of belief and knowledge is a "hard case" instance of our general abilities to render conceptions different from our own intelligible. We can accomplish this because there are at least *some* points of commensurability across cultures and, we should add, across scientific paradigms. However great our differences, we share much with those we seek to understand. In Putnam's words,

> Not only do we share objects and concepts with others . . . but also conceptions of the reasonable. . . . For the whole justification of an interpretive scheme . . . is that it renders the behavior of others at least minimally reasonable by *our* lights. However different our images of knowledge and conceptions of rationality, we share a huge fund of assumptions and beliefs about what is reasonable.[87]

Intelligible interpretations would be impossible otherwise.

Scientists *as* scientists share many concepts of what constitutes reasonable scientific discourse regardless of whether they embrace divergent paradigms or other orientations. As members of scientific communities scientists share interests in solving problems relevant to maintaining or expanding comprehension of that part of the natural world that concerns their specialty fields. Specific problems chosen and how they are formulated can vary considerably, both within and across scientific paradigms, but scientists as scientists share understandings about the kinds of problems that are *scientific* problems. It is reasonable for scientists to identify and address ambiguities or defects in existing evidence, in prevailing theories and interpretations, or in received methods and techniques. Scientists also recognize that scientifically reasonable problem-solving efforts should be pursued in accordance with values scientists share as scientists. Claims that possess qualities of empirical accuracy, consistency, quantitative precision, or explanatory power make sense; claims that are empirically inaccurate, inconsistent, quantitatively imprecise, or lacking explanatory strength can be dubbed scientifically unreasonable. These and other criteria of acceptability and conditions of exigence can vary considerably in specific applications both within and across paradigms and both within and across specific situations. Nevertheless, they are among the common, communal grounds of all reasonable scientific claim-making and evaluation.

Scientific discourse is discourse for and within a community that participates in and judges the situational reasonableness of claims made. The task of the scientific rhetor is to make discourse that implicitly or explicitly invokes concepts of reasonableness shared by members of the community to be addressed and to do so in ways that make situational sense. This requires that the rhetor engage grounds and principles that govern the informal logic of rhetoric in and about science.

SCIENTIFIC DISCOURSE IS INVENTED DISCOURSE

In an admirable and sweeping work of much significance the contemporary American philosopher Maurice A. Finocchiaro has described and illustrated a conception of rhetoric in science. His book-length study is titled *Galileo and the Art of Reasoning.* I wish to discuss Finocchiaro's work at some length because the study is important and because in so doing I can conveniently lay the groundwork for the markedly broader conception of the *inventional* rhetoric of science that I shall propose in chapters to follow.

Finocchiaro discusses two broad kinds of rhetoric that he finds at work in Galileo's *Dialogue*. One kind is a "purely rhetorical" dimension, a kind Finocchiaro distinguishes from the *Dialogue's* "intellectual content" or "actual substance."[88] The other kind of rhetoric has to do with the "verbal techniques of persuasion which operate on human feelings and emotions."[89] These two kinds of rhetoric clearly encompass a narrower body of theory and practice than I believe applies to scientific discourse, yet Finocchiaro has gone beyond purely managerial conceptions of rhetoric.

As Finocchiaro sees it, the first kind of rhetoric "refers to the appearances and pretensions found in verbal discourse, and the various *impressions* conveyed by it, as opposed to its substance."[90] Analyzing this sort of rhetoric allows discrimination of the *Dialogue's* "real" intentions from mere appearances created through use of "cosmetic verbal expressions of desires and intentions."[91] We learn to dismiss such impressions as (1) the *Dialogue* is an exercise in religious apologetics addressing the Church's anti-Copernican decree of 1616; (2) the *Dialogue* is an attempt at balanced presentation of arguments for and against Copernicanism; or (3) the *Dialogue* is an effort at rigorous demonstrative proof of the earth's motion. Instead, the *Dialogue* "is in reality a justification of Copernicanism."[92] Thus distinguished, we can better understand the rhetorical implications of Galileo's intention to achieve practical effects, such as persuading church officials of the truth of Copernicanism so as to cause the repeal of the condemnation of 1616.[93]

The second kind of rhetoric is concerned with a type of intellectual content, "but one that plays upon feelings and emotions, either directly and explicitly by verbal expressions that have the desired emotive effect, or else indirectly and implicitly by emphatic identification with what is explicitly said or done."[94] Finocchiaro's analysis of the special rhetorical content of the *Dialogue* identifies a grab bag of discursive strategies including, among others, fear appeals, emotional descriptions, imagery, *ad hominem, reductio ad absurdum,* ridicule, caricature.[95] Finocchiaro calls this dimension of Galileo's argument its rhetorical force:

> We have attempted to define a dimension of Galileo's argument distinct from its purely logical one and from its purely rhetorical one. We have found it in its combined emotional appeal and literary-aesthetic value, and we may label it its rhetorical *force.* This is distinct from the argument's logical structure . . . insofar as the latter refers to argument and appeal to evidence, whereas rhetorical force involves aesthetic images and appeal to emotions. This rhetorical force is also distinct from rhetorical appearance . . . insofar as the latter refers merely to communicative expressions, whereas the former pertains to persuasive effectiveness.[96]

How, if at all, rhetoric is related to argument and evidence, beyond "dressing" them and giving them emotional force, is not clear. Where and how Galileo *found* the arguments and evidence he chose to use is not made explicitly clear, but the implication is that this sort of discovery and selection ("rhetorical invention" in traditional terms) is the product of some sort of nonrhetorical, *logical* process.

Whenever Finocchiaro offers counsel on how best to analyze scientific discourse, he stresses that a logical inquiry must have priority over rhetorical analysis. Accordingly, he says that analysis should begin with a study of the logical dimension—a dimension conceived as going beyond orthodox formal logic to include "the activity of reasoning itself."[97] Finocchiaro maintains that one cannot even begin to understand the rhetoric of a scientific work without first exposing the underlying logical structure on which the rhetoric must be grounded.[98]

Finocchiaro and I agree that scientific discourse does not persuade solely by fulfilling logical criteria but, with other commentators on the rhetoric of science, I wish to go considerably further than he in characterizing the "rhetorical" dimensions of such discourse. Kuhn has contended that we cannot gain full understanding of the dynamics of scientific change without examining "the techniques of persuasive argumentation"[99] and without comprehending the "full force of rhetorically induced and professionally shared imperatives" that influence scientists' choices, which could not have been made according to the dictates of logic and experiment alone.[100] Clearly, Kuhn was thinking of "persuasion" and "rhetoric" in more sweeping terms than Finocchiaro. Other commentators have echoed Kuhn's thought, again using broader conceptions of rhetoric than Finocchiaro's. Weimer exclaimed that persuasive argumentation is "literally the essence of science,"[101] and he contended that scientific knowledge "is a matter of warranted assertion rather than proven or probable assertions."[102] Similarly, Simons argued that rhetoric is at work in scientific deliberations whenever there is "reason-giving activity in judgmental matters about which there can be no formal proof."[103] Overington emphasized that the construction of scientific knowledge involves argumentation before an audience.[104] Significantly, each of the last three writers is a practicing social scientist.

These and other writers draw attention to the persuasive character of scientific argumentation. Thus, while Finocchiaro views rhetoric as emotive and aesthetic dimensions added to an underlying logic that structures scientific argumentation, the broadened conceptions identify rhetoric with scientific argumentation before adjudicating audiences. This view implies but does not deal directly and specifically with principles concerning *how* arguments are constructed and evaluated *as science.*[105]

Recent work in the rhetoric of scientific and scholarly communities suggests that it will no longer suffice merely to claim that scientists, like other rhetors, choose among alternative discursive strategies when seeking to legitimate their claims; instead, we must begin to probe and extrapolate the underlying criteria which, for good or ill, constrain and structure the range of legitimate rhetorical choice-making within special communities of discourse.[106] Pursuing this line of thought requires addressing two traditional and fundamental questions about constructing any kind of rhetoric: (1) Can the choice of what to say in a given case be systematic even though not formally logical? (2) What are the criteria for these choices if they are not exclusively logical or impulsive?

My position is that the inventional principles that constrain and guide the creative and evaluative processes of scientific argumentation can be delineated by heuristically adjusting the perspective that general rhetorical theory supplies to the special problems, expectations, and values of scientific communities. I shall explain the evolved, special, inventional principles in terms sufficiently general to apply to all discourse conventionally called scientific, but these principles are nevertheless amenable to being "filled in" with conceptual and technical possibilities at varying levels of specificity within any scientific specialty. In the briefest and most general terms, choosing what to say *can* be systematic because when people try to prepare scientifically reasonable discourse that will be persuasive to a scientific community, they do, or they can do, what general rhetorical theory has described as (1) selecting fitting rhetorical ends, (2) locating the points of exigential clash, i.e., the *stases,* and (3) inspecting potentially relevant *topoi* to discover situationally appropriate lines of discussion. These choices are constrained by criteria which, although not formally logical, are by no means illogical; the method of choosing involves reasoning, but about how to create situationally reasonable claims rather than formally logical entailments. In the next three chapters I shall set forth the special inventional principles that constrain how scientists make these three kinds of substantive decisions—decisions that yield the rhetoric of science that we actually observe in scientific discussions.[107]

7

RHETORICAL INVENTION IN SCIENTIFIC DISCOURSE: Determining Rhetorical Ends

A comprehensive inventional theory of how scientific rhetoric is created must specify the "legitimate" ends of scientific discourse. In chapter 3 I identified four characteristics of rhetorical ends considered generally: (1) All rhetorical ends are persuasive ends. (2) Viable rhetorical ends must be possible given conventional norms embraced by situated audiences. (3) Rhetorical ends enable discrimination among relevant and irrelevant sayables within the intended situation. (4) Where audiences are heterogeneous, rhetorical aims can usually be achieved only after new evaluative norms or senses of exigence have been rhetorically created. These general principles apply also when a scientist chooses and develops reasonable rhetorical ends for discourse addressed to scientific audiences.

THE "REASONABLE" AS A SCIENTIFIC END

Viable rhetorical ends for scientific discourse must be appropriate to whatever conventional norms govern the interactions of scientific rhetors and audiences. At a minimum audiences must believe rhetors are pursuing objectives that are "scientific." There is a professional *sensus communis* that prescribes what it means to think and to act reasonably as a scientist. If a rhetor's apparent objectives violate normative principles of this *sensus communis,* that rhetor's ethos as a scientist is apt to be questioned and, perhaps, even censured. Scientific rhetors know they will be judged in this manner; they pay attention to building and protecting their scientific images as they select what to say and how to say it.

A community of scientists exists—a community committed to a superordinate set of values and objectives peculiar to doing science. Within the general scientific community smaller communities associated with scientific specialties also exist, but all share the superordinate normative principles that unify all scientists as scientists. The superordinate principles of science are seldom formally articulated. Their normative presence is clearly reflected in such familiar evaluations as "He's more journalist than scientist" or "That's religion, not science." As I have already pointed out, discourse that does not systematically attack scientific problems is not taken as scientifically reasonable. So, too, with other shared and overarching objectives.

All reasonable scientific thought, action, and language is expected to be dedicated toward maintaining and expanding a scientific community's comprehension of natural order, since bringing order to natural phenomena is the basis of all scientific endeavor. Science is beset with myriad ambiguities and uncertainties. Scientists pursue order by constructing or evaluating discourse that addresses those exigences. Scientific rhetors and audiences thus unite in discursively forging constructs that superimpose order on the contingencies of experience. Kuhn stressed the importance of this unifying, constraining process when he contended that scientific discourse takes place within paradigmatic frameworks.

Once adopted, a paradigm's orientation theoretically and instrumentally induces expectations about natural phenomena. These expectations, encouraged by a paradigm's categories, presumably will be satisfied when one experiences the phenomena. If these functions of a paradigm are not realized, the sense of order suffers and scientists begin to doubt either the discrepant claims or the paradigm itself.

Scientists also will be committed to scientific methods, as Kuhn indicates:

> There is another set of commitments without which no man is a scientist. The scientist must, for example, be concerned to understand the world and to extend the precision and scope with which it has been ordered. That commitment must, in turn, lead him to scrutinize, either for himself or through colleagues, some aspect of nature in great empirical detail. And, if that scrutiny displays pockets of apparent disorder, then these must challenge him to a new refinement of his observational techniques or to a further articulation of his theories.[1]

Commitment to ordering and commitment to doing so by scientific methods thus constitute two kinds of "logical" tests of rhetoric in science.

Scientific methods do not free scientists from burdens of decision making when they seek to impose order on pockets of disorder. Decision-making crises are exposed whenever scientists make conflicting arguments during professional deliberations. Decision-making difficulties are underscored further when scientists provide sociological examiners with inconsistent accounts about scientific deliberations. Latour and Woolgar explain:

> Growing sociological interest in the details of negotiation between scientists has revealed the unreliability of scientists' memories and the inconsistency of their accounts. Each scientist strives to get by amid a wealth of chaotic events. Every time he sets up an inscription device, he is aware of a massive background of noise and a multitude of parameters beyond his control; every time he reads *Science* or *Nature,* he is confronted by a volume of contradictory concepts, trivia, and errors; every time he participates in some controversy, he finds himself immersed in a storm of political passions. This background is everpresent, and it is only rarely that a pocket of stability emerges from it. The revelation of the diversity of accounts and inconsistency of scientific arguments should therefore come as no surprise: on the contrary, the emergence of an accepted fact is the rare event which should surprise us.[2]

Mary Hesse arrived at a similar conclusion regarding the importance of order in scientific activity. In her view the pragmatic criterion of "increasingly successful prediction and control of the environment" is the overriding value in natural science—one that transcends any particular conceptual scheme or theory or paradigm.[3] What Hesse calls "the pragmatic criterion" further confirms that there are normative principles that constrain reasonable scientific thought and action. An important distinction needs to be made, however, between Hesse's philosophical perspective and a rhetorical perspective on scientific discourse. Hesse seeks to identify a situationally transcendent regulatory principle that governs development of natural sciences.[4] I have no quarrel with the transcendental regulatory principle that Hesse offers as the pragmatic criterion, but if we look at everyday scientific rhetoric we must also attend to the many constraints on discourse that arise from immediate situations. The generalization I would offer—one that arches over all immediate, scientific, rhetorical situations—is that *to be judged reasonable and persuasive in any specific situation, scientific discourse must be perceived as identifying, modifying, or solving problems that bear on a specific scientific community's maintenance and expansion of their comprehension of natural order.* All scientific communities strive to solve technical exigences that, once solved, expand the community's comprehension of whatever field of the natural order the community professes. This is the *rhetorical,* pragmatic criterion governing the logic of reasonable scientific discourse. Some of its consequences are predictable.

If a rhetor's purpose appears to address problems of order and disorder within a field of knowledge, the audience of scientists from that field will judge the aim of the discourse reasonable. There are four kinds of ambiguities potentially persuasive rhetoric can address reasonably (I shall discuss them more fully in chapter 8). They are: (1) ambiguities about what does or does not *exist* in a field of natural phenomena; (2) ambiguities about the theoretical *meanings* of constructs and phenomena; (3) ambiguities about the value or *significance* of claims advanced; (4) ambiguities about scientific *actions* required to understand a field of natural phenomena systematically. Reasonable scientific discourse must address exigences associated with one or more of these sorts of ambiguity.

In science people can disagree about what ordering principles ought to be emphasized when addressing ambiguities in a field of natural phenomena. But even in disagreement most will accept Ziman's claim that "scientific theories appear as ordering principles that explain general classes of observational and experimental facts, including the taxonomies, 'laws,' causal chains and other empirical regularities that are discovered about such facts."[5] In Ziman's view scientific theories constitute "the vehicle by which a description of natural phenomena is expressed as scientific *knowledge.*"[6] "Reasonable" scientific work strives toward formulating, applying, and extending theories.

Most scientists will accept the idea that a sense of order is best derived from theories containing general and necessary propositions that help them explain, predict, and control natural phenomena.[7] In general, an explanation can be viewed as a reasonable argument that links what is supposed to be explained *(explanandum)* with a more general conceptual scheme *(explanans)*. Any principle that imposes order on experimental or observational facts is in some sense explanatory. In turn, general principles of classification, taxonomies, and "laws" are each explained by being incorporated within more general and comprehensive conceptual schemes.[8] Theories containing constructs with predictive power also contribute to scientists' comprehension of order, allowing them to anticipate probable occurrences and relationships among phenomena. Constructs with predictive power help scientists develop and impose order through controlled investigations. Control depends on the ability consciously to manipulate empirical regularities (captured by ordering theoretical constructs) in the interest of pursuing some desired goal.

Philosophers of science argue vigorously about the meanings and functions of explanation, prediction, and control. Rhetoricians are more interested in what counts as an explanation, a prediction, or a control in a given rhetorical engagement. Standards for what counts will vary among scientists and across rhetorical situations, but what remains constant is the general view that anything "reasonable" must contribute to a scientific community's maintained or expanded comprehension of natural order.

The uniquely scientific methods of dealing with ambiguities are experimentation and systematic observation. The scientist seeks to resolve problems of order by testing hypotheses about puzzling phenomena in the laboratory or the field. Problems of order amenable to such investigations are the primary concerns of scientific activity. True, scientists might pursue other kinds of problems which directly concern, say, theologians and philosophers. For example, scientists may interest themselves in the nature of reality or the origins and structure of the universe. But these are problems more to be reflected on than experimentally explored or systematically observed; they are not primary concerns of scientific activity. Thus, scientific rhetoric must speak the language of observation and experimentation if what is said is to be accepted as reasonable.

It thus becomes a logical requirement of scientific rhetoric that any order propounded must include the premise that its constructs can be related to empirical experience. The constituents of the proposed order must be grounded empirically and should not include, for example, supernatural constructs invoked in the writings of theologians or linguistic analyses such as philosophers often render. Persuasive rhetoric in science offers claims that allegedly cohere with conventionally accepted theories and constructs and that correspond with processes and events alleged to occur in the natural world. Behind such rhetoric is a tacit assumption that theoretical and experimental claims are in some way isomorphic with the natural world. This fact explains further why in scientific discourse there is so much emphasis on locating general and necessary propositions that allow for explanation, prediction, and control.

In summary, when scientists formulate and advance specific rhetorical ends, interested audiences will test whether the scientists' goals are to advance the conventional normative principle that all reasonable scientific thought, action, and language should contribute to a scientific community's comprehension of natural order. Reasonable rhetorical aims will be those that address exigences or ambiguities clouding comprehension of natural order. Such aims will engage issues about what does or does not exist, about the theoretical meaning of claims, about the scientific value of claims, or about the appropriate actions for scientifically investigating phenomena.

Despite the normative goal of reducing ambiguities, scientific rhetors still have to choose discursive strategies that will convince colleagues that specific aims and claims do further comprehension, and should be incorporated into the field's evolving literature as reasonable. These strategies may, of course, be selected consciously or according to habit originally inculcated by training in scientific methods. In either case the overarching norm of all scientific persuasion reveals relevant sayables and strategies that make logical sense. Put in rhetorical terms, the normative principle governing scientific logic directs attention toward certain *topoi* and away from others.

THE *TOPOI* OF SCIENTIFIC REASONABLENESS

The normative objective of scientific rhetoric implicates four major classes of rhetorical *topoi,* each of which suggests potentially persuasive themes for establishing a claim as scientifically reasonable. These *topoi,* or "headings" for persuasive arguments, are *problem-solution topoi, evaluative topoi, exemplary topoi,* and *topoi* concerning *scientific ethos.* I have introduced the ethos *topoi* in chapter 6 and will discuss the other three classes of *topoi* more fully in chapter 9, but my immediate concern is with how these *topoi* relate to establishing reasonable rhetorical ends for scientific discourse.

A scientist can claim that individual research advances the problem-solving interest of a scientific community. The problems addressed are frequently empirical in nature, but there can also be problems of "fit" among constructs or theories. For example, one of the problems raised about Ernst Cassirer's *Philosophy of Symbolic Forms* was whether or not his theory of how humans acquired powers of symbolic abstraction could be explained within the broader theory that natural selection governs the evolution of a species. This particular problem remains open for discussion for anyone interested in rendering Cassirer's theory scientifically persuasive. A scientist choosing to do this must engage themes that help establish this conceptual problem as a scientific problem and find grounds for arguing that a position should (or should not) be accepted because it will (or will not) solve that problem in a scientifically reasonable manner. The lines of thought which yield such specific arguments are what I call *problem-solution topoi.*

A second general source from which scientific rhetors may draw arguments arises from the values of science, or what Kuhn and others refer to as the "good reasons" for scientific choice. These accepted values include four especially important subtopics always at issue in scientific communication: accuracy, simplicity, scope, and consistency. If any claim or interpretation or presentation of data lacks these qualities it will not be maximally persuasive to informed scientists. Thus, to render a fresh idea persuasive one can show it is based on accurate data and methods of analysis, meets the test of parsimonious explanation, is generalizable, and is both internally consistent and consistent with the rest of existing knowledge. There are many specific ways of arguing that these and other values do or do not inform claims. Those arguments, as a class, are drawn from *evaluative topoi.*

A third broad source from which scientific rhetors can draw I call *exemplary topoi.* Specific forms of thought that exemplify ideas include analogies, metaphors, and examples. All can render claims persuasive. For example, H. H. Kelley has proposed that a layperson uses a "naïve version" of analysis of variance in attributing causes. One attraction of this theory is that human processes are seen as comparable to a scientifically recognized statistical procedure, analysis of variance, or ANOVA. Exemplary strategies such as this can further the premise that one's position restores or gives order to pockets of disorder in a particular phenomenal field. This is one appeal attribution theorists find in Kelley's Naïve ANOVA Model of attribution.

When there is ambiguity in the relationship between a rhetor's rhetorical objectives and the normative principle of maintaining or expanding a scientific community's comprehension of natural order, dispute is likely to arise about whether the rhetor is thinking and acting as a scientist or whether the discourse advanced ought to be addressed to scientific audiences at all. Generally, scientific audiences are likely to judge a discourse unreasonable if they think the rhetor lacks the requisite qualities of thought and conduct needed to adduce scientifically reasonable aims and claims. For instance, it might be argued that a scientist's apparent lack of objectivity and skepticism renders the technical reasonableness of claims suspect. Scientists must therefore display in their discourse qualities of thought and conduct which are believed to exemplify responsible and admirable scientific endeavor. More directly, scientific audiences will likely judge discourse unreasonable if they think the rhetor is posturing as a scientist while covertly pursuing extrascientific objectives. For instance, any member of an academic community has encountered charges that some members of the discipline are "popularizers" and not "serious scholars."[9] This kind of attack questions the individual's ethos as a member of a self-defined academic community. The individual's motives and objectives are seen as illegitimate, given the interests that the academic community professes. The professional ethos of a rhetor as a scientist can be rendered similarly ambiguous if there is reason to believe the rhetor's primary objectives involve such nonscientific pursuits as securing personal celebrity with lay audiences, generating wealth by serving powerful economic interests, achieving political aspirations, advancing religious principles, or perpetuating maverick beliefs that have supernatural or occult implications. To show that one's aims are consonant with the communally sanctioned principle of maintaining and expanding comprehension of natural order, a scientist may supplement technical arguments with explicit or implicit appeals to themes drawn from *topoi* related to scientific ethos.[10]

When there is ambiguity regarding the scientific ethos of rhetors, the rhetorical strategies employed against them are likely to focus on the reasonableness of the ends each individual pursues. In chapter 10 I show the rhetoric of so-called scientific creationism has failed as a rhetoric of science partly because the rhetorical aims of creationist discourse have seemed religious and therefore scientifically unreasonable. At this point I wish to look at two other illustrations of how scientific discourse is discredited by doubts concerning the reasonableness of the pursued rhetorical ends. The examples show that tests and defenses of reasonableness in science are frequently not instances of formal, logical analysis. Intentions, motives, purposes, can become very serious issues concerning what is reasonable in scientific disputes.[11]

The Case of the Parapsychologists

Parapsychologists have had difficulty establishing the scientific legitimacy of their objectives in part because of parapsychology's intellectual ancestry.[12] The immediate precursor to experimental parapsychology was spiritualism, the belief that the dead communicate with the living. Critics of parapsychology have used this history as a source for arguments against the scientific reasonableness of objectives pursued by parapsychologists.[13]

Psychologist E. G. Boring wrote, "It is quite clear that interest in parapsychology has been maintained by faith. People want to believe in an occult something."[14] In similar fashion T. S. Szasz contended that easy assimiliation of E.S.P. with the occult "is responsible for the obscurantism which pervades this area of inquiry and which makes its companionship unwarranted in the larger field of scientific disciplines."[15] Parapsychologists agree that association of parapsychology with the occult has been enormously persuasive against their specialty's scientific legitimacy. For instance, Beloff wrote; "Parapsychology has, all through its history, suffered from its fatal attraction for persons of unbalanced mind who seek in it their personal salvation."[16]

Efforts by parapsychologists to be scientific have been dismissed as either fraudulent or pretentious. Collins and Pinch point out that critics made their charges of fraud credible by demonstrating fraud's *possibility,* rather than by proving its *actuality.*[17] For instance, in an article in *Science,* Price directly accused parapsychologists Soal and Rhine of fraud—a charge he later retracted in a letter to the same journal. His original argumentation ran:

> We must recognize that we usually make a certain gross statistical error. When we consider the possibility of fraud, almost invariably we think of particular individuals and ask ourselves whether it is possible that this particular man, this professor X, could be dishonest. The probability seems small. But the procedure is incorrect. The correct procedure is to consider that we likely would not have heard of professor X at all except for his psychic findings. Accordingly, the probability of interest to us is, the probability of there having been anywhere in the world, among its more than 2 billion inhabitants, a few people with the desire and the ability artfully to produce false evidence for the supernatural.[18]

A more typical argument has been that parapsychologists' attempts to do scientific research are pretense, or that parapsychologists dress belief in the supernatural in scientific garb. Price contended that parapsychology, "although well camouflaged with some of the paraphernalia of science, still bears in abundance the markings of magic."[19] Similarly, in *Illusions and Delusions of the Supernatural and the Occult,* D. H. Rawcliffe said this about E.S.P. research: "To view the modern E.S.P. movement in perspective, one must realize that it is basically a cult—a cult of the supernatural in technical dress. The perpetuation of all such cults depends ultimately on irrational beliefs and the ignoring or 'explaining away' of rational criticism."[20]

All of the arguments reviewed so far indicate that critics of parapsychology believe that its objectives subvert the overarching standard for all reasonable scientific purposing: expanding or maintaining a scientific community's comprehension of natural order through using scientific methods. Parapsychologists' rhetorical aims become scientifically unreasonable when critics make them seem guilty by association with supernaturalists. Parapsychologists have difficulty escaping this line of attack when efforts to use scientific methods are dismissed as either fraudulent or thinly veiled attempts to advance beliefs in the supernatural through a misleading rhetoric of science.

Three significant *topoi* used to argue against parapsychology as reasonable scientific research are *explanatory power, experimental replication,* and *external consistency*. An argument drawn from the *topos* of explanatory power was used by Szasz:

> In the realm of psychical research, in spite of widespread interest and intensive effort over more than half a century, there is still nothing that would deserve to be called a theory, even by the most enthusiastic proponents of this work. This, in the writer's opinion, constitutes the most decisive factor which casts doubt upon the "reality" of the entire structure of parapsychology.[21]

Another argument against parapsychological research alleges such research cannot produce repeatable experiments.[22] Experimental replication, the *topos* of this argument, expresses an important value in assessing the scientific legitimacy of experimental claims. Critics claim parapsychological research does not exemplify this value. Crumbaugh wrote that establishing the repeatability of parapsychologists' experiments was "crucial and must be met before the great bulk of scientists will swing over to accept the E.S.P. hypothesis."[23] More directly, Cohen wrote in *The Nation,* "Obviously successful E.S.P. experiments are not repeatable and thus do not meet a basic requirement of all scientific experiments."[24]

Critics of parapsychology also draw arguments from the *topos* of external consistency. Such arguments urge that parapsychological research be rejected as scientifically unreasonable because its claims contradict what is already "known" in related scientific disciplines. For instance, parapsychological phenomena are alleged to contradict accepted laws of physics. As Allison explained it, parapsychological phenomena are "completely unrestricted by distance or any kind of physical shielding; the size of the material target has no effect; in the form of precognition, psi implies an effect preceding its cause in time."[25] The psychologist T. R. Willis made an argument drawn from the same *topos:*

> The conclusions of modern science are reached by strict logical proof, based on the cumulative results of numerous ad hoc

> observations and experiments reported in reputable scientific journals and confirmed by other scientific investigators: then and only then, can they be regarded as certain and decisively demonstrated. Once they have been finally established, any conjecture that conflicts with them, as all forms of so-called "extra-sensory perception" plainly must, can be confidently dismissed without more ado.[26]

The arguments I have cited derive from questions about the explanatory power of parapsychology, the replicability of its experiments, and its consistency with knowledge external to it. These examples are sufficient to establish that persuasion based on the three *topoi* cited does exist. The *topoi* I have presented identify recurring lines of thought for and against the worth of allegedly scientific claims. Moreover, the arguments illustrate that the logic of scientific discourse is *topical* rather than formal and that the *topoi* constitute both the headings and grounds for analysis of scientific discourse.

Parapsychologists have replied using arguments from *topoi* that scientific communities recognize and value. For example, they defend their claims by displaying their use of rigorous experimental controls and techniques, an argument that derives from the scientific value of *experimental competence*.[27] They argue that the anomalous character of parapsychological phenomena underscores the scientific importance of their research.[28] This kind of argument is drawn from the *topos* of *significant anomaly*. Anomalies purportedly show there is need for radical reconstruction of currently accepted scientific paradigms. Parapsychologists Schmeidler and McConnell used this type of argument when they wrote: "E.S.P. phenomena, such as telepathy and clairvoyance, are a type that has no place in the physical universe. . . . We are forced to conclude that the picture of the universe which present-day physicists have roughed out for us will have to be modified once again."[29]

The scientific reasonableness of rhetorical ends also can be attacked by questioning the relationship of a rhetor to a scientific community. If a rhetor's place in a scientific community is rendered ambiguous, professional ethos also becomes ambiguous. One typical strategy for making a rhetor's relationship with a recognized scientific community ambiguous is to claim the rhetor does not publish in sanctioned scientific journals. Comparatively few research reports on parapsychology have been published in such orthodox journals as *Science* or *Nature*. Instead, parapsychologists tend to publish in such journals as *Journal of Parapsychology* and *Journal of the American Society for Psychical Re-*

search. These journals are useful communication channels for parapsychologists but do not reach the broader scientific audiences whose favorable judgments are needed if parapsychology is to be considered a reasonable science. Of course, parapsychologists argue that the referees of orthodox journals are biased against them and repress their ideas. McConnell said:

> On the basis of experience in other fields I am convinced that the refereeing system frequently operates to suppress the publication of new and important material that happens to be personally distasteful to the referees to whom it is referred.[30]

In sum, to be persuasive, even within limited fields of science, publications must appear in sources recognized as reputable by "authorizing" audiences. Parapsychologists' problems are simply illustrative.[31] Virtually every branch of science has its set of highly reputable journals, publishers, and editors. What passes muster with these "gatekeepers" gives positive ethos to its author. Authors whose work is not accepted by, or is not submitted to, these guardians of "good science" lack persuasive ethos in or across scientific communities.

Another strategy for rendering a rhetor's place in a scientific community ambiguous is to reveal that the rhetor (or associates) had questionable sources of research funds. Allison has shown that research funds for parapsychological research come predominantly from lay sources, and sometimes with strings attached.[32] This presumably pulls parapsychological research in "unscientific" directions. For example, funds donated to gain "scientific demonstration of life after death" are alleged to have this effect. An endowed research professorship established at the University of Virginia was vulnerable to such criticism. The endowment stipulated that 50 percent of all research time be devoted to the question of "the survival of the human personality after death." Parapsychological associations, forced to compete with occult or spiritualist organizations for wealthy benefactors, further open their work to the charge of being unscientific. In one case a wealthy Arizona prospector left almost $300,000 to an unspecified institution that would try to find scientific proof for the existence of the human soul. The American Society for Psychical Research was awarded the money after a lengthy court battle.

Both argumentative strategies concerning a rhetor's place within a scientific community are based upon an important *topos* related to scientific ethos. The *communality topos* specifies that an important quality of scientific conduct involves participating actively in the intellectual life of a scientific community.[33] The two strategies just mentioned illustrate how it can be used to diminish perceptions of researchers' ethos as "scientific."

Scientific reasonableness also can be challenged by questioning directly the motives of researchers. In the case of parapsychology, parapsychological research received considerable attention from the media.[34] Though parapsychologists had difficulty addressing other scientists through recognized channels, they achieved celebrity with mass audiences. The usual battle cry was then raised by orthodox scientists: These people are showmen and not scientists. Rhine, a pioneer in experimental parapsychology, recognized how lay popularity helped to erode perceptions of parapsychology as legitimate science:

> The aggrandizing sensationalism which went on undaunted was a factor in generating the studied coolness to the work with psi. Many said as much. Parapsychology now belonged to the entertainer, the popular writer, the comic strip artists, and even to Broadway.[35]

Parapsychology is only one convenient example of how popularization potentially can diminish the perceived credibility of what aspiring scientists say. If, for example, the media publicize a medical "cure," its sponsors, if wise, will strongly assert the "cure's" limitations. If they do not, their ethos with the reputable medical community will suffer.

Such challenges to the legitimacy of scientists' motives are based on the important *topos* of *disinterestedness*. The line of thought is that credible scientists set aside self-interest in favor of securing community recognition for "making contributions to the development of the conceptual schemes which are of the essence in science."[36] Scientists' ethos is perceived favorably when they display habits of thought and conduct that seem disinterested or devoted to pursuit of "science for science's sake."[37] In contrast, when scientists appear to be professing technical claims as a means to securing personal celebrity with general audiences, their ethos with scientific audiences is likely to suffer. Such scientists will not seem sufficiently disinterested to be pursuing legitimate scientific objectives.

Scientists denounce with special severity the scientific reasonableness of research if members of the laity get involved in debate over scientific merit.[38] Those purporting scientists advancing the disputed claims are left exposed to charges of complicity with the laity in violating what I have already mentioned is a major taboo in scientific communities—seeking authorization of knowledge claims from audiences other than scientific audiences. This perception of scientists' conduct would be considered among the worst contraventions of the scientific virtue of *communality*.

Indeed, attacks on the scientific reasonableness of aims and claims would become even more strident when involved members of the laity are thought to believe in the "bugaboos of modern empirical science"—magic, witches, spiritualism, astrology, mysticism, or divination.[39] This fact led McConnell to warn members of the Parapsychological Association to be wary of the threat posed by "occult defilers of scientific parapsychology." He contended that "much of the reluctance of orthodox scientists to endorse extended support for E.S.P. research arises from their failure (and that of the lay press) to make a clear distinction between popular and scientific belief."[40]

Whatever the science, when this distinction becomes blurred, we can predict that rhetorical controversy will center on the scientific legitimacy of the ends discernible in the claims under evaluation. Not only does doing science involve making rhetoric, but within a scientific community aspiring scientists' apparent purposes are thought to be sound logical considerations in evaluating allegedly scientific claims. As a final illustration of the "logic" that operates when the legitimacy of scientific ends is established or disestablished, I offer Francine Patterson and Eugene Linden's book *The Education of Koko,* criticism of it by Thomas A. Sebeok, Professor of Anthropology, Linguistics, and Semantics at Indiana University, and Patterson's defense of her work.[41]

The Education of Koko

Sebeok is one of the most outspoken critics of language-acquisition studies of apes in general and of Patterson's work in particular. His arguments against Patterson's book exemplify the kinds of rhetorical strategies used against scientists when the scientific reasonableness of their objectives becomes questionable in the eyes of other scientists.

Sebeok attacked the scientific credibility of Patterson's claims for Koko's linguistic cleverness by arguing that those claims were externally inconsistent with what was known about language behavior in apes. He referred approvingly to Herbert S. Terrace's conclusion in his book, *Nim,* "that there is no evidence at all that apes can either generate or interpret sentences." Sebeok buttressed his argument by claiming Terrace's conclusion was externally consistent with the views of informed linguists and responsible ethologists:

> Terrace's results are . . . in perfect conformity with the long held judgment of informed linguists from Max Muller (1889) to Noam Chomsky. They accord equally well with the view of responsible ethologists, such as Konrad Lorenz, who declared,

> in 1978, "that syntactic language is based on a phylogenetic program evolved exclusively by humans," and that anthropoid apes ". . . give no indication of possessing syntactic language."[42]

The conclusion that Sebeok would have us draw from this passage is clear. Patterson's research claims are unreasonable because they are not grounded in, and consistent with, established research conclusions in respectable fields of scientific inquiry. Accordingly, as a scientist she can only be held in contempt as "uninformed" and "irresponsible."

Sebeok also questioned Patterson's experimental competence because she had not accounted for possibilities that the animal was behaving in response to cues given by experimenters, an experimental problem popularly known as the "Clever Hans phenomenon." According to Sebeok, this fallacy is one "by which Koko's entire ten-year curriculum has been arrantly nagridden." To support this contention Sebeok appealed to authority:

> The eminent Bristol neuropsychologist, Richard Gregory, also concluded, in 1981, that apes do not exhibit either "human language or intellectual ability," and wisely admonished: "There are so many experimental difficulties and possibilities of the animals picking up clues from the experimenters, given unwittingly, that extreme caution is essential."

Patterson's experimental competence was further put at issue when Sebeok challenged the interpretive claim that Koko was able to tell lies:

> Much is made of her aptitude for lying, which, according to the authors, "of course, is one of those behaviors that shows the power of language." Here, however, lurks a terminological confusion, one that, furthermore, begs the question. Many kinds of animals—the most remarkable case on record is that of the Arctic fox, *Alopex lagopus*—give, or give off, deceptive messages, in a word, prevaricate. But a lie must, by definition, be "stated," which Koko simply cannot do.

In another passage Sebeok seizes on an epigram—attributed to Koko—as exemplary of interpretive fallacies which, he alleges, riddle Patterson's research:

> A quotation attributed to Koko epigraphically opens the book, and, at the very same time, epitomizes its obstinate dottiness: "Fine animal gorilla"—this being her reply to the question, "Are you an animal or a person?" This exchange implies that

> Koko rediscovered the Linnaean system of classification and nomenclature. To the contrary, as Hediger has patiently explained, the string quoted is a purely human product that having been fed to the gorilla was regurgitated by her, and then reinterpreted as a novel sentence that seemingly originated in her mind.

In all of these criticisms Sebeok would have us conclude that Patterson's data are suspect because she does not understand the strategies for doing logically purposeful scientific work and so she sets nonscientific goals and makes nonscientific inferences. Her book should therefore be judged as marred by scientifically unreasonable aims and methods.

It is clear that criticisms drawn from the *topoi* of experimental competence and external consistency are used to imply that Patterson lacks technical virtuosity as a scientist. For these criticisms to diminish perceptions of Patterson's scientific ethos they must suggest she does not in some way think and act like scientists. The specific way of thinking implied here is indexed by the *topos* of *universality*. This *topos* recommends that we think about the behavior of scientists as involving willful and capable testing of claims against preestablished impersonal standards. Patterson fails to act like a "real" scientist because her claims do not meet two such criteria: (1) her externally inconsistent claims violate the intellectual consensus among scientists concerning what is accepted and rejected knowledge; and (2) her alleged experimental incompetence stifles her ability to find empirical grounds for her claims. We are left with the notion that Patterson's aims and claims become reasonable only if we replace "scientific" criteria with Patterson's comparatively idiosyncratic standards of judgment.

Sebeok continues his attack against Patterson's competence and purposes by using arguments drawn from the powerful *topos* of *skepticism*. He focuses on the fact that Patterson and Linden minimize the importance of emotional detachment and systematic doubt. He does this by associating them with proponents of a field whose scientific status and objectives have been called into question. They are "addicted to the use of ploys familiar from parapsychology, such as that the presence of a skeptic tends to ruin experimental results." Using the term "ploy" directly charges that the researchers do not put scientific purposing first. Among their ploys is their emphasis on emotional rapport between experimenters and animal subjects as a precondition for gaining successful experimental results. Sebeok quotes directly the authors' belief "that one cannot really understand the mental workings of other animals or bring them to the limits of their abilities unless one first has true rapport with them." He then asserts that the "obverse of this claim is that the intimacy between Patterson and her beloved Koko had hopelessly overclouded her scientific objectivity and judgment." In this way Sebeok invites us to conclude that Patterson's research claims must be technically unacceptable because her research resonates with an obstinate will to believe Koko makes and uses language. We are thus encouraged to judge Patterson's technical claims as scientifically unreasonable because she fails to display qualities of thought and conduct that reflect a skeptical outlook—a central virtue in conventional conceptions of scientific character.[43]

The reasonableness of a scientist's rhetorical objectives also can be rendered ambiguous by questioning that individual's relationship to the broader scientific community. To the extent that the rhetor's connection with the scientific community becomes confusing, so does the legitimacy of the rhetor's aims and claims.

Sebeok renders Patterson's place in the community of scientists questionable. He does so with three lines of argument based on the *topos* of *communality*. First, Patterson and her coauthor do not have memberships in any legitimate research institution, so neither is a "real" scientist. Patterson had written she was not able to analyze Koko's spoken language capacity in detail, despite an "enormous" amount of data collected. Sebeok offers this "translation": "In plain text, this citation means that since Miss Patterson's connection with Stanford University has been severed, she no longer enjoys free access to its computers." In other words, Patterson, "Koko's surrogate mother and pedagogue," has no scientific standing because she has lost the resources of a legitimizing institution. Sebeok attacks the scientific ethos of Patterson's coauthor, Eugene Linden, with even less subtlety. He depicts Linden as "a wrestler-turned-journalist, perhaps best known to the public for his *Apes, Men, and Language* (1974, 1981), surely the most gullible, as well as defensively emotion-laden, popular account of attempts at linguistic communication with any of our collateral ancestral species so far published." Linden, too, lacks connections with legitimizing research institutions; he is, at best, a popularizer of science—and not a very credible one at that.

In addition to arguing that Patterson and her coauthor do not belong to the scientific community by virtue of position, Sebeok "reads them out of" the community of scientists on grounds that Patterson was unable to secure ample public funds to support her research:

> While millions of dollars in federal funds were being squandered on the futile search for language in chimpanzees and

> orangutans, Patterson continued her work, without a proper institutional base, with the support of private sources, including a large, so-called "nonprofit" commercial enterprise, supplementing her income by minor grants from small Foundations.[44]

Sebeok asserts that this lack of public subsidy is "one respect" that "sharply" differentiates Patterson's research from other studies of apes' language capacities. The reader presumably is to understand that any minimally competent student of language in nonhuman primates would have had access to the all-too-available resources for doing the job "right," scientifically.

A third line of argument involves questioning Patterson's legitimacy as a scientist because she has failed to get her claims authorized by competent scientists. She had success reaching popular audiences through such channels as *National Geographic Magazine* and *Reader's Digest,*[45] but publication in orthodox sources was minimal, giving Sebeok warrant for asserting, "If Penny Patterson tried to publish in a scientific atmosphere, then she would be laughed out of court."[46] He attacks her for having a "warped perspective" and a "lack of receptivity to well-intentioned criticism." It is clear these are not the qualities of a scientist actively participating in the intellectual life of his or her community.

Sebeok also questions the authors' scientific ethos with remarks implying they are not sufficiently *disinterested* to be pursuing "legitimate" scientific objectives with the publication of their book. Sebeok offers innuendo about Linden's "real" interests in contrast with Patterson's emotional commitment to her research project: "Her co-author's stake in this enterprise—as well as, of course, his bond of personal relationship with the gorilla—is clearly of a different order." One implication drawn from this assertion is that Linden's motivations were not scientific, but perhaps were tied to making money. Sebeok attacks Patterson's motives more directly. He asserts that Patterson's motives for conducting research amount to little more than a "desperate reaching out for media recognition (of which this unfortunate book represents but one example)."

Sebeok's depiction of Patterson's work is intended to have us dismiss her claims and objectives as scientifically unreasonable. Whatever we may think of his charges and the rhetoric with which they are expressed, it is plain that he is making them on grounds that are scientifically logical, according to conventional understandings of scientific practices. We see again that in science special assumptions and their subordinate logical themes or *topoi* are constitutive of reasons for challenging or sustaining the scientific credibility of technical claims and claim-makers. We should notice that each theme Sebeok uses in criticism can also be the logical basis of claims *on behalf of* scientific credibility. Sebeok says Patterson is not connected with a legitimating research institution. The counterargument would assert one's association with legitimizing agencies. A person establishing scientific ethos would identify his or her scientific associations and his or her publications in scientifically refereed journals and books. These would display confirmations of that person's scientific purposes. Similar positive arguments could be developed from any of the *topoi* Sebeok used. Arching over all of his criticisms is the premise that science, properly conceived and executed, maintains and expands a scientific community's comprehension of natural order. Sebeok specifies what he alleges are reasons that Patterson's work does not and could not expand that comprehension. In sum, Sebeok contends that the authors' rhetoric is a ruse as science. He implies that is why their work receives adherence only from the laity.

It would be misleading not to mention that Patterson made arguments in her book that anticipated several of the rhetorical strategies Sebeok used. Her discourse indicated keen awareness that her professional ethos was at issue and likely to be challenged. She did not attempt to make positive, explicit claims about her place in a legitimating scientific community, but she focused on technical defenses of the legitimacy of her research claims and she sought to amplify qualities of thought and conduct that could enhance her scientific credibility. Among her means of persuasion were arguments drawn from the problem-solving *topoi* of *significant anomaly* and *experimental competence.* She bolstered her ethos with arguments implying that she possessed qualities of *individuality* and special research ability in establishing *experimental rapport* during her investigations.[47]

Patterson argued that objections to claims that nonhuman primates possess language capacity are symptomatic of a deeper intellectual revolution taking place within the behavioral sciences concerned with language. She was thus able to anticipate objections based on the *topos* of *external consistency* by fashioning arguments from the *topos* of *significant anomaly.* According to Patterson, after R. Allen Gardner and Beatrice T. Gardner published findings that the chimpanzee Washoe was able to use language, *Science* published rebuttals written by "the most distinguished names in the behavioral sciences." But Patterson claimed that the Gardners' "success" with Washoe presented "one of the most basic tenets of modern life" with an "anomaly"; there was now evidence that humans are not unique in their possession of language. She contended that the hostile reaction to anomalous claims indicated that something like Kuhn's notion of scientific revolution was taking place in studies of language acquisition in primates.[48]

Using the *topos* of significant anomaly to construct a revolutionary scenario for nonhuman primate language research has important implications for what constitutes a credible scientific ethos for researchers. Should audiences believe Patterson's claims, there would be no warrant for esteeming communality as a specific scientific virtue: there is no legitimizing, homogeneous community during periods of crisis. Judgments about what constitutes "legitimate" research institutions, "appropriate" sources of research funding, and "qualified" adjudicating audiences become partisan points of contention during "revolutionary" science. It is thus not a vice for scientists to neglect intellectual participation within the community in these ways. Instead, lack of such participation could actually be shaded into an admirable quality. The revolutionary scenario allowed Patterson to imply she possesses intellectual courage by following the Gardners: she boldly abandons outmoded but comfortable notions about language and pursues a pioneering line of investigation even when confronted with eminent hostile opinion. In showing unwillingness to buckle under to the authority of conventional knowledge and its distinguished spokespersons, Patterson portrays an admirable ethos insofar as one values the anti-authoritarian qualities of individuality.[49] Patterson's conflict with received knowledge and prevailing orthodoxy is thus raised to the heights of scientific virtue.

Patterson also made claims to experimental competence. In an effort to argue against critics' efforts to parallel the "Clever Hans" phenomenon with the results claimed about Koko's use of language, Patterson explained the experimental procedures that were used to avoid inadvertent cueing:

> I have checked my findings through double-blind testing, in which the ape can see the test object to be identified, but not the tester, and the tester can see the ape's response, but not the object. . . . This eliminates any possibility of cueing, and random changes in the order of the objects presented for identification prevent the ape from using a strategy like memorization to come up with the correct answers.[50]

She followed this explanation by stating that the rigor with which data were collected in the ape studies was not met in studies of language development in children. She contended that far from being a technically questionable research area, experimental designs used in ape studies had led to efforts to improve experimentation with children: "It is ironic that the collection of 'hard' data on language development in apes has spurred the search for better controls when studying language development in children."[51]

Throughout the book Patterson disengages scientific investigations of language from the commonplace association of scientific thought and conduct with making and assessing research claims in view of what Merton called "preestablished impersonal criteria." Linguistic scientists cannot turn to an established knowledge consensus for guidance because "there is very little that can be said about language today that is not open to question or controversy."[52] Nor can they ground claims restfully in rigorously empirical criteria. At one point, Patterson contends,

> There is much about language that does not lend itself to reduction to statistics and hard data, and some linguists have recently reacted against the rigid, formalized treatment of language. When we speak with each other, we are not isolated by double-blind procedures. Indeed, a good deal of our comprehension of the spoken message comes from a perfectly natural "Clever Hans" appreciation of the nonlinguistic cues to meaning of the message. This is not to justify vagueness but to illustrate that it is very difficult to speak with any confidence of "facts" about language.[53]

She then presents several anecdotes allegedly showing that Koko's complex linguistic behaviors cannot be demonstrated by controlled experiments and strict interpretations of data, but can be understood fully only through assessing the animal's intentions within specific behavioral contexts.[54] Accordingly, she chose not to construct an experiment for documenting Koko's alleged use of humor, because it "would be an enormously complex and perhaps impossible task." She suggests instead that "we interpret the meaning of this anecdote [Koko's joking] through devices that are an ordinary part of understanding any message, but that do not fall neatly into the empirical method of any scientific discipline."[55] These claims work implicitly toward minimizing universality as a source of persuasive themes for establishing or challenging scientists' ethos.

Patterson does not argue that scientists should set aside empirical methodology entirely, but she claims one must supplement rigid experimental work with interpretive case studies in order to obtain reliable knowledge about the full range of nonhuman primate linguistic capacity.[56] Case studies require that researchers possess special interpretive sensibilities, which allow them to achieve what I shall call experimental rapport with subject beasts. This means that adequate scientific investigation of any sentient being's behavior requires a special capacity for securing emotional attachments with animal subjects. Patterson argues specifically that only those researchers able to establish true rapport with the animal being studied will meet a necessary condition for gleaning positive evidence of language acquisition in nonhuman primates; in contrast, those who adhere rigidly to objective, experimental stances toward animal subjects will confound possibilities for discovering significant language use. The experimental rapport argument can thus, at one stroke, amplify Patterson's and like-minded colleagues' scientific ethos and minimize the credibility of critics insofar as the "best" research encourages development of rapport between scientist and animal:

> In none of these cases did the experimenter [Terrace or Premack] allow himself to develop a true, close rapport with his chimp. This was justified in the laudable name of objectivity, but given the sensitivity of the animals involved—Koko's signing is affected by even slight disruptions in her routine—it is hard not to wonder whether the different conclusions about ape language abilities reached by these scientists ultimately trace back to the different relationships between experimenters and subjects and to the persistence that has marked the efforts of those of us who have established close rapport with our subjects. If this is the case, I am reaffirmed in my belief that one cannot really understand the mental workings of other animals or bring them to the limits of their abilities unless one first has true rapport with them. Even the critics admit this possibility. What they fail to see is that the problem really is a misunderstanding of the purpose of language. Once that misunderstanding is straightened out and we accept language as a communicative behavior, the evidence of Koko's abilities is compelling for those who want to see it.[57]

This argument is based on the *topos* of particularity, which suggests research claims can be legitimately assessed on the basis of personal criteria including a researcher's established reputation, experience, and technical skill. In this instance a central and indelibly personal criterion is a researcher's special ability for establishing emotional rapport with subject animals.

I am not interested here in whether Patterson or Sebeok "won" their "debate." The points I wish to make are (1) that purporting scientists do, in fact, have a set of acknowledged *topoi* that identify ways in which the scientific legitimacy of claims and actions can be reasonably or logically judged; (2) that the *topoi* used in scientific discourse index specific lines of argument that may be used either to affirm or deny that a communicator's purposes (and practices) are scientific; (3) that purporting scientists can disagree regarding the relevance of specific *topoi* or specific applications of a single *topos* given variations in their objectives, interests, and overall perceptions of situational contingencies; and (4) that scientific discourse appears "reasonable" or "logical" only if ideas are drawn from among these *topoi* and are relevant to these communally agreed-upon patterns of thought. If a purporting scientist deviates from standard applications of conventional scientific *topoi* or challenges the legitimacy of the *topoi* themselves, the deviation or challenge must be strongly supported. A scientific rhetor choosing to make revolutionary arguments based on the significant anomaly *topos* rather than submit claims to tests of external consistency should expect to be challenged. So too should a rhetor extolling qualities of individuality over communality, or ability to achieve experimental rapport rather than objectivity or skepticism. Whether Patterson defended her use of these arguments successfully is not the point here. My point is that she or any other purporting scientist who makes unconventional or revolutionary arguments assumes a special, scientifically "logical" burden of proof.

CONCLUSION

The critiques of parapsychology and Patterson's research show us much of what scientific rhetors must do in order to convince scientific audiences they are pursuing reasonable rhetorical ends. The reasonableness of rhetors' rhetorical ends can become a matter for debate in at least three ways. First, rhetorical ends can be charged directly with being extrinsic to a scientific community's commitment to expanding and maintaining comprehension of natural order. Efforts to discredit the claims of parapsychologists illustrated this fact.

Second, when a rhetor's rhetorical aims fall within the legitimate purview of making rhetoric about science, those aims can still be discredited on technical grounds. It may be argued that a certain body of scientific discourse fails to contribute to the community's comprehension of natural order for one or more of the following reasons: the claims offered have no or limited explanatory power, or they defy experimental replication, or the claims are externally inconsistent with received experimental claims and theories, or their support was derived in experimentally incompetent ways.

Third, the reasonableness of a rhetor's rhetoric about science can be diminished or enhanced through attacks or encomiums concerning a rhetor's "scientific" qualities of thought and conduct. For instance, consider how the communality *topos* can be used to generate questions regarding a rhetor's standing within the pertinent scientific community. How seriously should scientists treat the knowledge claims of someone who works outside recognized seats of scientific investigation? Is it a sign of a scientist's reasonableness that he or she is clearly a part of an establishment? Can discourse addressed to the laity actually be scientific? Or should a scientist consistently address only peers on scientific subjects? Should serious scientists devote time to evaluating research done by persons who do not formally associate themselves with established kinds of research organizations, with traditional sources of funding for scientific work, and with established channels for authorizing and disseminating scientific knowledge for the field? On the other hand, is it a sign of scientific reasonableness that a rhetor confines himself or herself to the research patterns and resources common to "our" community? Questions like these *do* arise in scientific discussions, as the examples discussed clearly show. The questions identify argumentative strategies from which specific charges of unreasonableness and claims of reasonableness can be developed. Other *topoi* related to scientific ethos can yield additional possible strategies. Collectively, these *topoi* constitute the "places" where scientists find ways of evaluating whether rhetors think and act like scientists. The results have direct implications for diminishing or enhancing the apparent reasonableness of scientists' discursive aims.

It is also the case that every evaluation of a scientist's rhetorical objectives is inevitably an assessment of his or her ethos as a scientist. This is true in all rhetoric. Any rhetor's apparent aims make up part of the data on which his or her character and wisdom are judged. What is less often recognized is that when serious controversies arise between or among scientists, the ethos of the disputants becomes a major factor in the scientific community's evaluations of claims and counterclaims. The important rhetorical point is that *persons,* as well as data and logic, influence scientists' judgments, even though there is a myth that scientific discussion disregards personal qualities. Whether a purporting scientist displays personal qualities befitting *scientists* when making rhetoric is always one determinant of how the scientist will be judged; and in sharp controversies past credibility, method, and manner often become decisive factors in the community's ultimate decisions.

8

RHETORICAL INVENTION IN SCIENTIFIC DISCOURSE: Deciding What the Issues Are

As Howard Gardner put it, science never yields "a completely correct and final answer." Scientific activity always admits some degree of ambiguity and uncertainty:

> There is progress and regress, fit and lack of fit, but never discovery of the Rosetta stone, the single key to a set of interlocking issues. This has been true at the most sophisticated levels of physics and chemistry. It is all the more true—one might say, it is all too true—in the social and behavioral sciences.[1]

Since there is no Rosetta stone, scientists must choose where, in existing knowledge, they will try to insert their own scientific comments. Science, no less than any other human endeavor, demands that its practitioners make responsible choices. Scientists—not nature—choose which problems to work on and how to formulate those problems; scientists decide which avenues of investigation are likely to yield solutions; and they decide what claims will seem to constitute progress or regress in their community's comprehension of natural order.

To choose responsibly when dealing with science is to make choices consonant with the logic specially valued by scientists. The choice of where and how to insert claims or other comments into the ongoing accumulation of scientific understanding is a *rhetorical* choice and is judged situationally and scientifically "logical" or "illogical." Scientists must choose the issues they will address, and they need to show their peers that issues addressed are logically significant given the present state of scientific knowledge. Put differently, they must establish for themselves and others the relevance of their problems and proposed solutions whenever that relevance is not self-evident. In this chapter I shall outline the framework of logical options within which scientific rhetors decide about situational and scientific relevance.

STASIS PROCEDURES FOR SCIENTIFIC DISCOURSE

Science does not advance in the presence of ambiguities and differences; the objective of science is to certify knowledge as both comprehensive and univocal. For this reason any point of disagreement or uncertainty—any "gap" in certifiable knowledge—constitutes a technical exigence that needs to be altered if possible. Technical exigences give rise to multiple yet finite issues that could be addressed, for there is never only one way to redress a scientific difference, ambiguity, or gap. General rhetorical theory posits that there are logical procedures by which one can reasonably weigh the pertinence of issues and choose which ones deserve address and in what order. Scientific discourse reflects procedures that identify the issues that arise within the logical framework of situated, scientific discussion. In the language of traditional rhetoric, scientific *stases,* or points at issue, are displayed in ongoing scientific discourse. The purpose of this chapter is to set forth the character and logic of those specifically scientific *stases.*

Scientific discourse shows there are four general kinds of "stoppage" or "stands." These I shall call *superior stases.* Superior *stases* identify arguable points concerning the four grand functions of doing science: adducing evidence, interpreting constructs and information, evaluating the scientific significance of matters discussed, and applying scientific methods. Discussants must agree on such matters or scientific activity cannot expand comprehension of natural order. Within the realm of each of these broad classes of issues or *stases,* subordinate problems have to be settled. There are questions about such matters as the availability, the meanings, and the usefulness of evidence, constructs, judgments, and procedures. These points of detail I shall call *subordinate stases.*

As is shown in Figure 8.1, I observe four superior stases and sixteen subordinate stases being used in actual contemporary discussions of science.

Superior stases are those that arise from recurring kinds of impediments to a scientific community's comprehension of natural order. All reasonable rhetorical purposing in science addresses and tries to modify one or more of these kinds of exigence or ambiguity. *Evidential stases* emerge from ambiguities about what does and does not exist in a phenomenal domain. *Interpretive stases* spring from ambiguities concerning the meanings of constructs and phenomena. *Evaluative stases* arise from ambiguities concerning the values attributed to experimental, theoretical, or methodological claims. *Methodological stases* arise when there are ambiguities about procedures for scientific action.

Figure 8.1 Rhetorical Stasis Procedures of Scientific Discourse

	Superior Stases			
Subordinate Stases	**Evidential**	**Interpretive**	**Evaluative**	**Methodological**
Conjectural	Is there scientific evidence for claim x?	Is there a scientifically meaningful construct for interpreting evidence?	Is claim x scientifically significant?	Is procedure x a viable scientific procedure in this case?
Definitional	What does the evidence mean?	What does construct y mean?	What does value z mean?	What does it mean to apply procedure x correctly?
Qualitative	Which empirical judgments are warranted by available evidence?	Which interpretive applications of construct y are more meaningful?	Which claims are more significant, given value z?	Which investigations exemplify appropriate applications of procedure x?
Translative	Which evidence more reliably grounds claims about what does and does not exist?	Which scientific constructs are more meaningful?	Which scientific values are more significant?	Which procedures more usefully guide scientific actions?

Scientific rhetors seek to frame their communicative objectives and their subordinate claims to address one or more of these kinds of exigences because scientific rhetoric always attempts to reduce ambiguities. The exigences may already exist in the audience's perceptions, or the rhetor may want to bring them to the audience's attention. In either case, if discourse is to seem reasonable as science, what a rhetor says must address some evidential, interpretive, evaluative, or methodological problem that impedes the community's comprehension of natural order. Anything that does not address such problems is scientifically irrelevant for the community.

Points at issue in science always concern one or more of problems about *existence, meaning, value,* and *action.* All four kinds of problems may be raised in discussion of a single topic. Consider a circumstance where scientists ask themselves whether they have enough knowledge to render a relatively firm judgment on some scientific matter. One scientist may argue that the relevant phenomena are not understood with adequate specificity. Another may argue that the principles used in explaining the phenomena are more crucial than the specificity of the data. Still another may question whether the data or judgments at hand have sufficient scientific value to justify the inquiry; and someone else may argue that what really is needed is not more of the present kinds of data but a new kind of strategy in collecting data. None of these lines of thought would be scientifically irrelevant; but were someone to inject notions about whether it is convenient to make the present inquiry, that claim might be humanly relevant but would certainly not be considered scientifically relevant.

Even when scientists share perceptions that a certain ambiguity needs clarification, they can disagree about how the problem should be formulated or about which specific issues need to be resolved to solve the problem. At this point what I am calling *subordinate stases* come into play. There are four kinds of clash that can occur whether the superior *stasis* point is evidential, interpretive, evaluative, or methodological. These subordinate issues fall into one of four classes: *conjectural, definitional, qualitative,* or *translative.*

When discourse takes place in an evidential framework, attention focuses on exigences or ambiguities about what does and does not exist in the natural world. Here discussion is about establishing whether some object, process, or event exists or has existed and whether what is known can be known more precisely, in quantitative and qualitative terms.[2] Scientifically reasonable answers to questions of existence require empirical evidence and related judgments.

An ethologist doing field work on the behavioral patterns of a specific species may encounter a behavioral display that apparently has never before been recorded. This scientist must then try to make sense of this otherwise ambiguous event and reach some judgment about whether the striking behavior witnessed indicates an unforeseen pattern of behavior in the species. The ethologist might continue field observations hoping to discern evidence of recurrent displays of the odd behavior, but he or she is limited because there is no clear way for the observer to induce the strange behavior through experimental intervention. In contrast, experimental observation of phenomena is distinguished by the scientist's deliberate and controlled intervention in the natural world. Evidential exigences then arise concerning the scientists' decisions about how to elicit experimental information in novel and fruitful ways.

I bring together under the single class of *evidential* exigences those problems scientists encounter whenever issues arise concerning the scientific relevance and situational appropriateness of empirical evidence and related judgments used to address questions about existence. Scientists' evidence is always potentially at issue, regardless of whether information constituting that evidence was produced in natural or experimental settings. When an exigence is an evidential exigence, scientific rhetors must deal with one or more of four subordinate *stases*. These four kinds of subissues define the possible "points of entry" for scientific discussions of evidential matters. If one knows and checks them, one can weigh which line(s) of attack on a problem will most reasonably advance answers to empirical questions of existence.

Conjectural stasis occurs in evidential scientific discourse whenever there is ambiguity about the availability or reliability of evidence. The specific kinds of conjectural issues are too numerous to catalog exhaustively, but the following kinds of potential problems are representative: Is there empirical evidence that supports or resists confirmation of a scientist's favored hypothesis? If there is evidence, is it reliable? Do supporting experimental data really recommend the phenomenon as postulated, or are the reported phenomena artifacts of experimental apparatus or statistical procedures used to "make" the data? Have results been reproduced by additional experiments, showing their reliability? Do the postulates of a favored received theory encourage scientists to deny or to accept the possibility that the data exist as reported?

As an example of conjectural "stoppage" in evidential discourse about the existence of phenomena, consider Collins's examination of Joseph Weber's claim to have found a new natural phenomenon: high fluxes of gravity waves.[3] Much of that controversy turned on what constituted a "working" gravitational wave detector. Weber built the first such device, and in 1969 he claimed to have detected high flux gravitational waves with it. However, Weber's supporting data were attacked as spurious. One of his critics brought discussion to a point of conjectural clash when he wrote: "The Weber group has published no credible evidence at all for their claims of detection of gravitational radiation."[4] An anonymous scientist interviewed by Collins suggested Weber's questionable statistical techniques could have "fudged" results: "By massaging data again and again, knowing what you want for an answer, you can increase the apparent statistical significance of any bump. I'm pretty sure he could get there out of pure noise."[5] Both comments take a conjectural stand against the reliability of data adduced for Weber's empirical claim.

We also find scientists challenging claims made at the fringes of orthodox science on the grounds the claims lack credible empirical evidence. Westrum makes the point that scientists can reject the existence of anomalies either because everything scientists know theoretically militates against an anomaly's existence or because strong evidence for the anomaly is absent.[6] If there are good theoretical reasons for doubting, say, the existence of UFOs, those reasons will be applied in challenging the reliability of evidence (usually eyewitness sightings) adduced for their existence. The phenomenon cannot be established without the strongest possible evidence—for example, reliable photographs, or finding a disabled craft. The general question for deciding this kind of issue becomes: Is there or is there not *scientific* evidence for claim X?

When there is ambiguity about what the available evidence means, the issue becomes *definitional*. An issue of definition arises concerning evidence when, for example, experimental results are puzzling, or accuracy in categorizing data is challenged, or an observer is thought to have mistaken cause for correlation or the reverse. Problems that occur when scientists try to fit individual cases or sets of data into existing taxonomies are among the most representative cases of definitional *stasis*. Paleontologists, especially, struggle with classification problems when they unearth novel fossils. In each such case definitional *stasis* occurs regarding evidence when there is ambiguity about the meaning of individual specimens or sets of data. The problem does not lie in the taxonomic principles and classification schemas themselves.[7]

Definitional stoppages came into focus among paleontologists during a controversy about how to fit conodonts accurately into received taxonomic categories. According to Gould, this task was quite problematic:

> Conodonts are evidently the only hard parts . . . of an otherwise soft-bodied creature. But what kind of animal, and how can you tell from a few separated toothlike structures? When conondonts were known only as isolated, disarticulated elements—the situation from their discovery in 1856 until 1934—we had almost no anchor for any sensible opinion, and speculation ran rampant. Conodonts were placed in almost

> every major group of the plant and animal kingdoms, from support structures for algae to copulatory organs of nematode worms. The most common opinions cast them as jaw elements either of annelid worms or of fishes.[8]

Scientific discussions from 1856 to 1934 focused on a definitional *stasis*. The ambiguity concerned what those fossils meant within the frameworks of taxonomic categories and received knowledge. This point of clash, like all other definitional *stases* about evidence, raised the general scientific question: What does the evidence mean?

Qualitative stases emerge in evidential discourse whenever there are questions about how available evidence can be applied in reaching conclusions about the existential status of phenomena. In experimental papers scientists make final empirical assessments of interpreted experimental data. For instance, conclusions can be used to qualify interpreted results as, say, proving empirical existence or merely suggesting its probability. Scientists might qualify their data by extrapolating some linkages with models or other constructs used to account for quantitative of qualitative features of the phenomena. Examination of review articles assessing the state of research on almost any experimental problem will almost surely show some scientists arguing that there is sufficient evidence for warranting a phenomenon as fact. Others will attack this judgment as premature, contending that the balance of experimental evidence warrants only more or better investigations. Still other scientists are apt to cite some especially convincing experiment and urge that its results alone warrant making a final judgment. This kind of clash concerns which empirical judgments have the best quality, given available evidence. All qualitative clashes and obstacles about evidential problems require an answer to this scientific question: Which empirical judgments about phenomena are warranted by available scientific evidence?

Controversies in the memory-transfer field supply excellent illustrations of scientists raising qualitative points of stoppage to rebut judgments that the transfer phenomenon exists. Proponents of the memory-transfer phenomenon assumed that if a laboratory animal were trained to perform a particular task, the animal's brain would code this behavior in molecules specific to the task. To demonstrate that memory is coded in transferable molecules, special experiments were designed. Scientists trained animals to a task, sacrificed them, and removed their brains. They sought to extract chemicals from the brains by either preparing homogenates of entire brains or applying purification procedures to extract RNA from the brains. Scientists injected these preparations into untrained animals.

When the untrained recipients performed the original task better than control groups, this was taken by some as evidence that the memory-transfer phenomenon exists.

The experimental case for the existence of memory transfer was dealt a serious rhetorical challenge by a letter published in *Science*. It was signed by twenty-three scientists representing several laboratories located at highly prestigious institutions. The central claim advanced in this piece was that eighteen independent experimental efforts to corroborate positive findings for memory transfer had failed.[9] This claim raised and addressed a qualitative stasis in evidential scientific discourse. It bypassed the conjectural issue concerning the availability of credible scientific findings and the definitional issue by assuming that experimental results were interpreted accurately as yielding evidence against the existence of the memory-transfer phenomenon.[10]

Proceeding in this way left the audience with the qualitative question of whether this new negative evidence was sufficient to warrant the empirical judgment that the memory-transfer phenomenon did *not* exist. In an interview one scientist made it clear that he was convinced this judgment was warranted: "If those guys couldn't get it, well, there was no point in me trying." On the basis of this and other evidence, one commentator on the memory-transfer controversy concluded that the letter was "received by many as a blanket refutation."[11]

Translative stases or stoppages occur when there are problems about which evidence provides the better basis for making existential claims about phenomena. There is this kind of *stasis* in social scientific investigations of the influence of pornography on those who consume it. Some social scientists contend that reliable claims are possible only with statistical evidence that depicts behavioral trends over long periods of time. Others opt for detailed case histories of violent criminals and pursue correlations between antisocial acts and use of pornography. The issue dividing these two classes of investigators is the translative question: Which evidence yields the more reliable grounds for claims about what does and does not exist?

The *interpretive* frame of scientific discourse arises from ambiguities in the meanings of theoretical constructs used to account for data. This kind of *stasis* becomes prominent when scientists accept sets of data as facts but have difficulty deciding what theoretical constructs or models accommodate them.[12] At times, points of incompatibility between different theoretical explanations indicate deep conceptual differences about what meanings to attach to data. When problems are interpretive, rhetorical efforts at advocating, clarifying, and applying theoretical constructs can clash at any of the following four points of *stasis*.

In interpretive discourse there are often questions about whether there *are* theoretical constructs that can meaningfully explain or classify recalcitrant data. We then have *conjectural stases*. Arguments about whether anomalies pose significant threats to received theoretical constructs and, if so, whether those constructs should be rearticulated or replaced address such a conjectural *stasis*.

In invertebrate paleontology there is a category called "problematica," which contains well-preserved fossils of unknown zoological affinity.[13] Should paleontologists try to make the case that available taxonomic categories are insufficient to account for fossils and that they should be tossed into this class, the point at issue would be whether a new category or schema is needed. Such conjectural *stases* in interpretation occur concerning the availability of meaningful categories.

One example of conjectural ambiguities is found in Linnaeus' trouble describing the coot. He defined the bird generically as having lobed toes but later came to know and identify species of coots with smooth toes. When the empirical accuracy of a taxonomical definition becomes the focal question, not the reliability of evidence or how evidence should be classified in accepted taxonomies, we have conjectural *stasis* about interpretation: the interpretive fertility of available constructs is the locus of ambiguity and the focus for discussion. In Linnaeus's case there was clear need for a new generic category that incorporated the characteristics of all species of coots.[14]

Similarly, in 1983 Briggs, Clarkson, and Aldridge provided detailed descriptions of a novel conodont animal fossil, showing that despite similarities with two available taxonomic categories (i.e., chordates and chaetognaths) the animal did not accurately fit within either one. They went on to argue that a new phylum needed to be formulated for conodonts. In their view the full meaning of the fossil could not be drawn out without categories allowing a better taxonomic fit.[15] These illustrations show conjectural *stasis* occurs when dispute comes to be about whether data defy categorization because of problems with received principles of ordering. All such points of clash give rise to this scientific question: Is there or is there not a scientifically meaningful construct for interpreting (i.e., identifying, classifying, explaining) evidence?

Definitional stases occur in interpretive scientific discourse when the meanings of constructs are at issue. Scientists can define theoretical and hypothetical constructs differently. This blocks the progress of scientific discourse. For example, a common criticism of social scientific research on televised violence is that there are scores of operational definitions but a paucity of conceptual definitions for aggression. The question becomes: What does "aggression" mean? Similarly, empirical research on source credibility can be challenged on the basis of this *stasis* point. Do researchers share conceptual meanings when they claim that variables contributing to credibility include competence, personality, and moral character?

Definitional conflict also occurs when principles used to develop classification schemes are questioned or are otherwise confused or ambiguous. As a case in point, definitional *stasis* between geneticists and paleontologists about the meaning of "species" is a crucial consideration in the controversy concerning the theory of punctuated equilibria. Paleontologists distinguish a fossil's species according to morphological differences, while geneticists distinguish the species of living organisms by their inability to interbreed.[16] Observe also how an ethologist struggled to work out the meaning of the term "learning," a term that confounded discussion about memory-transfer research using planarians:

> What we started with were definitions with which I could agree but as the discussion has gone on, I have become more confused. As Denny described response, I was in complete agreement; as he described learning, I was in complete agreement as an ethologist working with animals in the field. I have worked with animals in their natural habitats and the definition of learning, as I understood Denny, fits in beautifully not only with mammals but with birds as well. If we are to have a definition of learning we are going to have to take notice of evolution and what animals do under natural circumstances. I would say that learning is a modification of the stimulus-response complex which results in a new favorable adaptation of the whole animal to its environment.[17]

Driver was responding to a definitional *stasis* in interpretation. Like all fitting responses to definitional clashes, his discussion tries to answer a question characteristic of all such disputes: What does scientific construct y mean?

Qualitative stases concern interpretive exigences when the central ambiguity is how scientific constructs should be applied in accounting for data or in interpreting subordinate concepts within a conceptual scheme. This kind of issue does not concern conjectural questions about whether there is a scientific construct. Neither does it involve definitional discussion about the meaning of a construct. In many situations scientists assume those questions have been answered, or they bypass them and make arguments about the quality of alternative ways of applying a particular construct. For example, in response to Denny's presentation of a learning model, Jensen raised the question of whether the model could properly be applied to the behaviors of nonmammals.

> I would like to emphasize that the characteristics outlined by Denny apply to mammalian learning. One wonders if these same stages would apply in describing the learning of a shark, for instance. I am now studying an animal which is a primitive vertebrate, the myxinoid fish, and, as far as we can determine, it demonstrates no exploratory behavior. This might present some difficulties for the model. Whether the animal can learn or not is yet to be determined.[18]

The passage shows a concern with the range and the quality of applications for the model.

As with other kinds of *stases,* qualitative *stases* point to a particular kind of scientific question. Consider some specific examples. Does resource mobilization theory account for motivations that lead to the inception of religious movements? Can social learning theory be applied to the massive attitude shifts involved in religious conversions? Can the considerable evidence of people's reluctance to change be accommodated by dissonance theory? These and similar questions pose qualitative *stases.*[19] They all take the general form of this central question: Which interpretive applications of construct y are meaningful?

Translative stases arise in interpretive scientific discourse when doubts are raised about which construct should be used in interpreting the meanings of claims. The following specific questions create translative *stases*: Does resource mobilization theory or deprivation theory provide the better explanation of the motives that underlie the inception of cult movements? Does social judgment theory or dissonance theory provide the better understanding of evidence that people resist change? Later in this chapter I shall cite arguments concerning the issue: Does "learning" or "sensitization" provide the better account of experimentally induced behavioral changes in planarian worms? All translative *stases* in interpretation raise the general scientific question: Which scientific constructs are more meaningful?

Evaluative scientific discourse occurs when the significance of claims is questioned. In Thomas Sebeok's review of Patterson's work he challenged the significance of Patterson's observations for linguistic science. What is at issue in such cases is whether data or claims have value for the scientific community in which they are proposed. Questions raised include whether the data or claims are relevant as introduced, are clear, are novel, or are otherwise useful for the scientific specialty with which the audience is concerned. Comparative values also may be argued. Rhetors may disagree over which data or claims are more or less fruitful. Whether intrinsic or comparative values are ambiguous, the resulting discussion will be evaluative rather than, say, interpretive or evidential. Evaluative discussion can focus on any of the four secondary *stases*.

Conjectural stases occur in the evaluative frame of scientific discourse when one asks whether a claim has scientific value at all. The point to be settled is whether the intrinsic value of an idea justifies its consideration in the rhetorical situation. For instance, Sebeok contended Patterson's claims were inaccurate and inconsistent with received linguistic knowledge. For these and other reasons he inferred that her claims did not possess scientific value. The central question to be decided in *stases* of this sort is: Is claim x scientifically significant?

Definitional stases are points of clash where there is uncertainty about what a scientific value means in a particular circumstance. When a social scientist argues that the normal margin for error of a public opinion poll is smaller (or greater) than is required for sound judgment of tendencies in opinion formation, the issue raised is definitional and is raised in order to allow a sound ultimate evaluative judgment. The question is what accuracy means in this case. To take another instance, scientists frequently ask what it means in a given case to say a theory or data claim is consistent with other theories and claims. When this kind of evaluative problem arises, the question to be decided is: What does scientific value *z* mean in our present circumstance?

Qualitative stases occur during evaluation when one asks how specific scientific values apply in comparative assessments of data or theoretical or methodological claims. What is asked is not whether a claim possesses a scientific value such as accuracy; the issues raised are how and to what degree claims put forward are significant when accuracy is comparatively applied. For example, Kuhn said of the oxygen and phlogiston theories, "One theory . . . matched experience better in one area, the other in another. To choose between them on the basis of accuracy, a scientist would need to describe the area in which accuracy was more significant."[20] The point is generalizable. Scientists argue over which claims are more significant or less significant. When they do, the central question is: Which claims are more significant, given value z?

Translative stases emerge if there is a clash about which scientific value should be used in judging significance. As Kuhn put it, "The relative weight placed on different values can play a decisive role in individual choice."[21] One rhetor may argue that a particular explanation is parsimonious, but someone else may contend that the real issue is not parsimony but empirical accuracy. Someone else may believe that empirical accuracy is surely desirable, but accuracy depends on proper controls. Still another scientist may raise the question of generalizability. In each case the underlying issue is, Which scientific values are more and less significant here?

When scientific discussion focuses on ambiguities about making and applying procedues and techniques, the issues become *methodological*. Typically, procedures scientists choose and describe in research reports are those the authors think will meet the audience's approval.[22] Procedures reported are likely to be standardized ways of getting at the facts or of calculating data. If the audience understands and endorses the procedures, they are more likely to judge the authors and their claims favorably; but the spell of controlled objectivity cast by following the conventions of scientific writing can be snapped at any time if someone raises questions about materials and equipment used, procedures selected, or applications of standard procedures. Methodological concerns are so central to scientific activity that resolution of evidential, interpretive, or evaluative *stases* frequently requires that these points of *stasis* be reframed as problems of methodology. Focus on this type of ambiguity distinguishes methodological scientific discussion from other kinds.[23]

Conjectural stases occur im methodological discourse when the point at issue concerns the scientific legitimacy of a method or procedure. For instance, in projecting election outcomes from poll data it was assumed for a time that "undecideds" would divide evenly between two candidates. Polling scientists eventually convinced each other that this was a mistaken assumption and ought to be replaced by more subtle polling of undecideds. New methods may be challenged as not having been sufficiently tested or as being imprecise. Unconventional methods may be rejected outright as pseudoscience. For example, Sebeok charged Patterson's practice of establishing rapport with experimental subjects was unconventional procedure. For Sebeok and others this procedure vitiated the research as science. Old procedures and methods may be attacked as antiquated. For example, development of x-ray motion pictures has led speech scientists to discount and finally to discredit data from the former practice of taking stroboscopic photographs of the action of vocal folds. In any such cases the central question is: Is procedure x a viable scientific procedure in this case?

Definitional stases occur in methodological discourse when the question concerns what it means to apply a scientific method or procedure competently. There can be ambiguity about how to apply statistical techniques, about how to select and make laboratory apparatus, about how to decide what supplies are needed, or about the proper sequences of steps in experimentation. When scientists follow the conventional practice of publishing a "good set of specifications" for repeating experiments, they are trying to minimize this kind of ambiguity as a source for conflict.[24] All *stases* about what to do when one adopts a technique or procedure occur at the definitional level. The central point for decision is: What does it mean to apply procedure x correctly in this case?

Qualitative stases occur in methodological discourse when the problem is which claims exemplify the best uses of an accepted scientific procedure. The crux of this issue is not the use of method per se, but whether the use of method has been of sufficient quality. For example, anthropologists frequently discuss whose applications of field study methods are most rigorous. A similar question might be: Have the most rigorous statistical tests been applied? Alternative data may be claimed to exemplify more competent applications of accepted mathematical procedures. Patterson's defense of her research involves qualitative argument about method when she claims that investigations of language acquisition in champanzees and gorillas have methodologically superior quality compared to studies examining language acquisition in children. Patterson invites her readers to conclude that language acquisition research with apes is based on more rigorous and tightly controlled procedures than those used in widely accepted research on children.[25] Clashes of this kind raise the general scientific question: Which investigations exemplify appropriate applications of procedure x?

Translative stases arise within methodological discourse when there is a problem about which among alternative methods or procedures ought to be used. Questions like the following are raised: Is controlled laboratory study superior to field study for present purposes? Is the orthodox and essentially Linnaean classification a better approach to plant taxonomy than experimental approaches?[26] Which method of biological classification is best: phenetics, cladistics, or evolutionary classification?[27] Scientific rhetors clash over such questions when there is ambiguity about whether some single methodology or combination or methodologies can sufficiently guide scientific actions. In *stases* like this the central question to be decided is: Which method or procedure more usefully guides scientific actions in the present case?

Traditional *stasis* points were discovered in discourse and, once discovered, were used analytically and heuristically as ways of discovering what to say in a given case. Roman rhetors needed to find points of potential dispute and to examine their relevance to different sorts of legal cases. In practice the classical *stasis* procedures were ways of logically discriminating relevant from irrelevant materials for legal and deliberative pleading. I contend that scientific discourse exhibits a comparable, comprehensive, kind of analysis that is logical for science. However informally recognized it may be, scientists have their own logical system that defines what it is possible and fitting to say in specific scientific situations. On the basis of that belief I wish now to consider the heuristic uses of that system for creating and evaluating scientific argument.

STASIS ANALYSIS: AN EPISODE FROM THE MEMORY-TRANSFER CONTROVERSY

Scientists *do* constantly make systematic choices of issues and argue about them. That should be clear from previous discussion and examples. There is no evidence, however, that the systemic character of scientific options has been noticed in modern times. I suggest that if the recurring points of decision are recognized as specifying the "logic" or argumentative give-and-take, they can give the same inventional guidance to practicing scientists that comparable analyses have given legal and political rhetors. Others may wish to rename and perhaps reorganize or add to the catalog of scientific *stases* I have extracted from contemporary scientific persuasion, but I would make three major points about any organization of the *stasis* points of scientific discourse: (1) The *stases* identify the major stands or points of clash to which scientific rhetoric is prominently addressed. (2) The *stases* of scientific discourse derive logically from the value system shared by those who do science. (3) The *stasis* points of science are systemic; they express scientific priorities because they derive from the systemic standards of scientific decision making and because they reflect the finite body of themes and purposes that are "authorized" by scientific communities.

If one wants to know how rhetorically intelligent a scientist's choices have been, one needs to analyze what that scientist's options for taking a stand were and which of them would have seemed most reasonable to those engaged in scientific discussion. Scientists choose, consciously or not, what is to be argued and what is not worth arguing. Whether the scientists realize it or not, there are only so many moves one can make at any given point in argument. One may try to accomplish the specific rhetorical maneuvers a given exigence allows, or one may try to reframe the exigence so as to allow for other kinds of maneuvers. The *superior stases* I have isolated identify four classes of scientific exigences, with each class possessing a limited number of subordinate rhetorical possibilities identified by the *subordinate stasis* questions I have presented.

I shall illustrate how *stasis* analysis yields critical insight into rhetorical choices by returning to the memory-transfer controversy and analyzing a paper presented by James V. McConnell, a psychologist then working at the Mental Health Research Institute in the University of Michigan, plus a body of ensuing discussion. McConnell's paper, "The Biochemistry of Memory," was published in the proceedings of a symposium on the chemistry of learning, sponsored by the American Institute of Biological Sciences, at Michigan State University, September 7–10, 1966.[28] A major portion of that conference was devoted to discussing research with flatworms. Such invertebrate research was the earliest kind of investigation into the molecular bases of memory. A central hypothesis at the conference inheres in the question: Are memories stored in molecules?

McConnell was known for his pioneering and controversial experimental efforts to test this potentially revolutionary hypothesis. Essentially he performed three kinds of experiments. In 1955 McConnell (and Thompson) claimed they had classically conditioned a species of planarian, or flatworm, to associate light with shocks. The claim that flatworms could *learn* was novel but did not create the kind of stir McConnell's later experiments did.[29]

With two students, Jacobson and Kimble, McConnell conducted regeneration experiments. They claimed that their findings indicated memories might be stored chemically in planarians. When cut into pieces, planarians can regenerate into viable worms. In the experiments worms were trained, cut in half, allowed to regenerate, and then tested to see how much of the initial training the regenerated worms retained. Results indicated that worms regenerated from brainless tail sections retained as much of the conditioning as those regenerated from head sections. On the basis of the evidence, McConnell and his colleagues claimed that memories were stored chemically throughout the planarians' bodies and were not localized in the brains.[30]

McConnell reasoned that if memories were coded in molecules, it should be possible to extract some of those molecules from trained animals and inject them into naïve animals, thereby chemically transferring to the naïve animals what the trained animals had learned. The question was: How should this be done? McConnell found his answer in the special cannibalistic qualities of the particular species of planarian used in his experiments. He explained the technique used to test the memory-transfer hypothesis thus:

> After trying several techniques that failed to work, we hit upon the idea of feeding trained planarians to untrained cannibals. So we conditioned some worms to the usual criterion, then cut them in half, and fed them to hungry cannibals. We also chopped up untrained planarians and fed them to another group of cannibals. Then, a day or so later, all of the cannibals were trained. We found that the cannibals that had eaten the trained worms were, from the first trials, significantly more responsive to the conditioned stimulus (light) than were the cannibals that had eaten control animals. Indeed, the control cannibals were scarcely different in their subsequent behavior than were planarians that had not been allowed to cannibalize at all.[31]

Scientists challenged McConnell's research program, and others like it, by making three kinds of refutative arguments.[32] They claimed that replicative results failed to corroborate positive experimental findings. For instance, some critics argued experimental results failed to indicate that training induced behavioral changes at all. Without such evidence there was no reason to test the specific memory-transfer hypothesis.[33]

A second kind of refutative argument involved advancing the claim that results like those of the Thompson and McConnell experiment could be explained as something other than learning through classical conditioning. Critics opted for alternative explanations like "sensitization" or "pseudoconditioning."[34] For example, the sensitization explanation contended the scrunching-up behavior performed by worms in response to light (conditioned stimulus) without the shock (unconditioned stimulus) was not due to their learning to associate the two stimuli, but was better explained by physiological changes leading to increased sensitivity to all stimuli rather than just the light.[35] Pseudoconditioning occurs when conditioned and unconditioned stimuli are administered in a manner that disallows associative learning but could increase the response to one of the stimuli.[36]

The "null" and the "nonlearning" arguments led to considerable disputation involving a third kind of argument, which Travis called "arguments about the nature of replication."[37] Both nonlearning and null arguments led to demands for clearer and more complete specifications of experimental details so that scientists could know what decisions they must make to replicate experiments that yielded positive results, including McConnell's. There was considerable ambiguity about what counted as a "competent" planarian experiment, and scientists seized upon this as a persuasive resource both for arguing against and for defending McConnell's claims.

The decline of interest in planarian research was not entirely due to the rhetorical force of the counterarguments. A series of announcements declared that research with mammals yielded positive evidence of the memory-transfer phenomenon. The claim that rats could be classically conditioned—could be said to learn—was far less controversial than similar claims about worms. Seemingly as a result of this "bias," scientists had already started shifting the field of experimental investigation from "worm running" to "rat running" by the time McConnell was addressing the symposium. I shall not review these developments because my concern is to illustrate how analyzing *stases* sheds light on discourse concerning McConnell's planarian research.

McConnell was inserting his claims into a complex rhetorical situation in which scientists could not agree on the specific questions they ought to be asking about planarian research in order to further understanding about the biochemical bases of learning. Conference participants had already published knowledge claims about technical exigences in the area, and so had some stake in how research problems should ultimately be framed. This is especially true about McConnell. He was perceived by many as a controversial and scientifically unorthodox figure.[38] His revolutionary claims about the specific memory-transfer hypothesis were thought to conflict with received electrophysiological theories of memory. Some speculated, however, that should his claims be positively established, he could very well become a candidate for the Nobel prize. With attention now shifting from planarian to mammalian studies of memory transfer, we can anticipate that McConnell would seek to control the course of scientific discussion by influencing decisions about which problems scientists ought to be pursuing to further comprehension about the biochemistry of memory.

McConnell indicated at the opening of his paper that he was keenly aware of arguments in opposition to his published views. He insinuated the reason skeptics did not accept his claims was that they preferred the comfort of conventional notions over truthful interpretive statements about potentially revolutionary findings. Had he been wiser, he said, he would have more effectively practiced the "art of public relations" to make claims more palatable:

> If I could change things, I would not give up the worms, and I would not give up any of the experiments I have done (although many could have been better performed), but I do believe that those of us early in the field made some classic mistakes in the fine art of public relations. We were, in our green and callow days, pretty naive about the sort of reactions we should elicit from our colleagues. Instead of glossing over the possible significance of our findings, we stupidly said out loud, for everyone to hear, that we found some pretty intriguing things in the laboratory. Telling the truth about such findings is probably a basic mistake under the circumstances; had we glossed over the possible importance of the regeneration and cannibalism studies, had we neglected to state our own interpretations of these results, we should not have got into as many public battles with noted scientists and we should surely have had less trouble getting research funds.[39]

If we ask what *stasis* points McConnell thought he could most reasonably address in this controversial rhetorical situation, we see that his presentation worked toward establishing three main contentions. First, a belief in the chemical-transfer phenomenon was warranted by available experimental evidence. Second, "learning" more meaningfully accounts for the results of conditioning studies than "sensitivity." Third, it is scientifically more important to investigate the transfer phenomenon than to debate whether planarians learn or undergo "mere sensitization." These were the stands McConnell inserted into the ongoing discussion about the chemical planarian studies.

McConnell's three major contentions dealt respectively with the superior *stases* of evidence, interpretation, and evaluation. They were offered as clarifying answers to questions about empirical existence, scientific meaning, and scientific value. Clarifying such ambiguities is consonant with the controlling normative principle of all reasonable rhetorical purposing when making scientific discourse. Certainly McConnell's formulations of the exigences were debatable, but they were not unreasonable insofar as research on the chemical bases of learning and memory was concerned.

McConnell's first main contention was an answer to the scientific question that arose from a *qualitative stasis* about *evidential* exigences. The issue dealt with what empirical judgments were warranted by available evidence. In McConnell's specific case the question was: Is the claim that acquired behaviors can be transferred chemically from one planarian to another warranted by available experimental evidence? To answer this qualitative question about experimental data, McConnell needed to present a full case for the affirmative answer he wanted to give. Presenting a comprehensive case involved confronting *definitional* and *conjectural* as well as qualitative issues. His rhetorical situation required a comprehensive case because the exigence concerned what constituted "warranting" where chemical transference occurs. This remained uncertain in the minds of many who would judge McConnell's discourse. Undecided scientists had conjectural and definitional, as well as qualitative, doubts. Some also were concerned about the reliability of McConnell's evidence. There was definitional confusion about whether behavioral changes were due to learning (i.e., conditioning), sensitization, or something else. Such conjectural and definitional problems needed to be addressed in order to make a scientifically reasonable response to the qualitative issue of sufficiency of the evidence.

McConnell's discussion of evidential exigences at first centered on definitional issues concerning the meaning of three bodies of planarian research evidence: Do the regeneration data mean conditioned memories are stored chemically in planarians? Do the cannibalism data mean conditioned memories are transferred chemically from one planarian to another? Do the RNA data mean RNA is a memory molecule? McConnell attempted to eliminate these questions as points of definitional stoppage.

As part of his attempt to remove the first point of definitional stoppage, McConnell launched an attack against the critics. He contended that rather than giving careful consideration to experimental evidence derived from his regeneration studies, some critics raised conjectural objections about the reliability of his evidence due to theoretical bias against McConnell's definitional claim that planarians could learn (i.e., be classically conditioned) and that what they learn is stored chemically throughout their bodies. He, a psychologist, took zoologists to task with special severity:

> Now, if Jacobson, Kimble, and I had been smart, we would have scrapped that [regeneration] study immediately and performed another quite different one in which we showed that sensitization, or perhaps even habituation, could be retained in regenerating pieces of planarians. That finding would have been exciting enough, and perhaps people would have believed us since nobody cares much about sensitization or habituation anyhow. Instead, we had the audacity to say that we had data suggesting that memories could be stored throughout the planarian's body, and the battle was on. One noted zoologist told me, rocking back and forth in her chair as she spoke, that she could believe that planarians could learn, but she could not believe that they could remember for more than 5 minutes. When I asked her what data led her to this belief, she merely snorted at me and said that she needed no experiments to tell her obvious truths like that. Another zoologist told me our data could not possibly be true because, if the tail of a planarian could remember, a zoologist would surely have discovered the phenomenon years ago. And yet a third informed me that he would not believe our data even if he repeated the experiment in his own laboratory and got the same results as we did. Now, this third zoologist was an exceptionally truthful man; for, a year or so later, he did in fact repeat the regeneration work, he got positive results, and he still refuses to believe the findings.[40]

McConnell similarly accused those who were skeptical about the cannibalism studies of being blind to the idea that learned behavior could be transferred:

> Again, had we been wise, we would have scrapped our initial findings and performed a study in which we fed sensitized planarians to cannibals and showed that sensitization transferred cannibalistically. I do believe that most people would have believed us, for few people have bothered to think through just what sensitization is and how long-term sensitization could be mediated chemically in any organism. Once we had convinced people that sensitization could be passed along to another animal, we might have tried habituation, as Westerman did subsequently (1963a and b). Once we had shown that this too worked and after we had given the scientific world

> time to digest these findings, we might have been able to announce that learning could be transferred cannibalistically without raising nearly as much dust as we did by starting with the learning study first.[41]

McConnell was claiming that some skeptics rejected his data because they were too biased theoretically to accept his assessments of what the data meant. In the language of *stasis* analysis, McConnell was asserting that it was not logical to take *conjectural* stands against the reliability of evidence because one does not accept *definitional* efforts to assess evidential meaning. The genuine point of stoppage for McConnell is definitional. To make a reasonable and comprehensive response to definitional stands McConnell needed to show (1) that his regeneration and cannibalism data were reliable, and (2) that they meant what he said they meant. He responded fittingly by touching on conjectural concerns about reliability by referring to experimental replications that yielded corroborating data. He tried to resolve the definitional obstacles by showing how explaining data as evidence of "sensitization" was not situationally reasonable.

McConnell sought to establish the reliability of evidence for retention of learning by referring to other experiments showing similar results. He mentioned investigations yielding both positive and negative evidence for retention of learning in head and tail regenerates but concluded, "An analysis of these various studies suggests that the transfer occurred whenever the maze habit was firmly established in donor animals."[42] He strengthened his definitional stand by referring to other studies whose data could not "reasonably" be assessed as sensitization: "The maze studies by Ernhart and Sherrick (1959) and by several others of us, as well as Westerman's experiment, make it difficult indeed to raise the specter of simple sensitization as an explanation of the phenomenon of retention of prior training in the planarian."[43] He followed the same pattern of argumentation for establishing the reliability and meaning of cannibalism data. In this case he referred to an experimental replication that allegedly yielded evidence for chemical transfer of learning even stronger than his own.[44]

McConnell confronted an especially complex rhetorical task when he turned to the RNA studies. He wanted to claim that data suggested RNA was the memory molecule, but he had to admit the poor quality of an experiment producing these interpreted findings. To minimize conjectural doubts about the reliability of RNA data, he once again referred to replicating experiments:

> Building on [the] early work by Corning and John, my students and I undertook to test the RNA hypothesis in a slightly different way. . . . We made every mistake in the book; our only excuse for performing such a clumsy experiment is that it was the first of its kind. Nonetheless, we did get evidence that some kind of transfer mediated by an extract had taken place (Zelman *et al.*, 1963). Luckily for us, other groups were able to replicate our early work using much more rigidly controlled conditions.[45]

McConnell then addressed the definitional obstacle by interpreting results from "the most noteworthy of these replications" as showing "RNA taken from well-conditioned animals does cause a transfer" and that "one would hardly expect any transfer of sensitivity via RNA."[46]

McConnell had so far addressed conjectural and definitional issues concerning evidence. He then shifted attention to what in his view was the sole experimental obstacle blocking the *qualitative* judgment that some form of learning transfers chemically from one animal to another.[47] Hartry, Keith-Lee, and Morton's experiment was "the most quoted but least understood experiment in the planarian literature."[48] In an extended critique of that experiment McConnell concurred with the experimenters' interpretation that the data failed to show transfer of conditioning. However, he claimed conditioning did not transfer because Hartry, et al. proceeded in a manner that was questionable as a test for transfer of conditioning:

> We are still left with an interesting problem though: Why did not the conditioning transfer? One possible reason could be that, instead of feeding their cannibals chunks of tissue, as was done in all other cannibalism studies, these authors ground up the experienced worms into very tiny pieces. Now, whenever one grinds up living tissue, one releases ribonuclease; hence, when one extracts RNA from tissue, one always makes sure that the tissue is extremely cold prior to homogenation, since cold helps neutralize ribonuclease. If RNA is indeed the transfer molecule, it might have been badly degraded before the cannibals got around to eating the finely ground up tissue provided them by Hartry, Keith-Lee, and Morton. . . . But there is a much more obvious fault with this otherwise excellent study. . . . When I remind you that Hartry, Keith-Lee, and Morton used massed training techniques and when I tell you that they used clean troughs rather than slimed troughs, perhaps you will begin to suspect what my criticism is. Frankly, I

> do not think the animals in group 1 were well conditioned at all; rather, I think they were mostly sensitized, for one would expect the animals in this group to reach criterion the second day in 40 trials or even much less if true conditioning had taken place on day 1. If only a small amount of conditioning took place on day 1, but a lot of sensitization took place, one would expect pretty much the results shown in this table. The fact that all the rest of us who obtained positive transfer results used spaced training and slimed troughs supplies some measure of support for my interpretation of the Hartry, Keith-Lee, and Morton results.[49]

McConnell had directed audience attention to concern about the experiment's *methodological* quality. He claimed the experiment tested for sensitization, not conditioning.[50] In other words, the study supplied data that should not count as disconfirming transfer of learning. To complete this part of his case McConnell then discussed at length an experiment he and a colleague had conducted with the Hartry et al. study in mind. This study used the "right" methods and so was of high quality. The resulting data were more readily explained as evidence of conditioning than as evidence of sensitization.[51] It was thus that McConnell allegedly removed the last definitional obstacle blocking authorization of the qualitative judgment that evidence is sufficient to warrant the existence of some form of chemical learning transfer.

It is likely that critical scientists would not be satisfied with this rhetorical maneuver. The shift from definitional *stases* about evidence to qualitative *stasis* about methodology would be convincing only if there was some consensus about what experimental guidelines should be followed to replicate McConnell's work. This was not the case. Some critics asserted McConnell's work was beset with definitional problems concerning experimental procedures. I shall discuss this point at greater length later, but it should be understood that ambiguities about procedures and standards for replicating planarian experiments remained within the rhetorical situation.

McConnell's argumentative pattern of addressing definitional obstacles about evidence could be convincing to scientists willing to accept his definition of learning; but again, definition was what skeptics had challenged when they said the evidence was evidence of something other than learning. McConnell nonetheless made a scientifically reasonable response to this outstanding issue insofar as he addressed (1) *conjectural* obstacles concerning reliability of evidence by referring to experimental replications that yielded corroborating data, and (2) *definitional* obstacles about the meaning of data by arguing that alternative constructs like "sensitization" did not explain the evidence. This follows the rhetorical logic of science, which prescribes that once the existence of reliable evidence is granted, the next *stasis* point becomes, What do these data mean? However, until this question is addressed and an answer agreed to, it is premature to address qualitative questions about empirical existence—a point McConnell seemed to recognize in his rhetorical situation.

McConnell did not move on to make the qualitative claim about sufficiency of evidence for establishing the empirical existence of chemical transfer of some form of learning. It is clear from his wording that he thought it would be scientifically logical to do so, but situational constraints were such that skeptics would not likely believe that his critique and replication of the Hartry et al. experiment sufficiently warranted this qualitative claim about evidence. McConnell implied full authorization was not a problem of sufficient evidence, but involved ambiguities in the meanings and applications of *interpretive* constructs used to dismiss evidence of conditioning and its transfer in planarian research. He contended those arguing for transfer of conditioning had stated their interpretive framework clearly—briefly reviewing key supporting tenets of the memory-transfer hypothesis—and suggested that those using sensitization to explain acquired behavioral changes do the same.[52] However, before shifting completely to interpretive exigences, he proceeded to make a bottom-line, qualitative claim about what empirical judgment the evidence does warrant regardless of how conflicting interpretations of evidence are ultimately settled. He first conceded: "Much of this [memory-transfer] explanation is highly theoretical and subject to change without prior notice." He then asserted:

> But I submit that the phenomenon of transfer itself is on rather good grounds. We may continue to argue about what it is that gets transferred and what the mediator is, but by now we all should be willing to admit that something is passing from one animal to another chemically that affects the recipient's behavior. When you stop and think about it, that is a rather remarkable accomplishment in and of itself, a phenomenon that was not even dreamed about some ten years ago. Even in the event that it turns out that learning per se cannot be transferred, the cannibalism and RNA injection studies in planarians may still have made an interesting contribution to our understanding of animal behavior.[53]

McConnell was showing considerable rhetorical skill in adducing this claim. Implying that he had addressed definitional obstacles sufficiently to warrant the qualitative claim that some form of learning transfer exists, he explicitly advanced instead the qualitative claim that experimental evidence a fortiori warrants the existence of chemical transfer of *something*. It was thus that McConnell attempted to avert the definitional obstacles he addressed but could not hope to remove in this rhetorical situation.

When McConnell turned to interpretive problems concerning how constructs like sensitivity and learning should be understood and applied, he began by touching on definitional issues. He admitted the two constructs are difficult to define precisely and distinguish operationally, but he nonetheless offered some guiding principles: "Sensitization is an increase in responsivity to all incoming stimuli, not due to any pairing of the conditioning and the unconditioned stimuli, while learning is an increase in responsivity that is due to the pairing of the conditioning and unconditioned stimuli."[54] Assuming proper controls for experimental discrimination between the two ideas, McConnell asserted that there are theories of learning (e.g., classical conditioning theory) that allow qualitatively rich applications to experimental settings but implied there is no comparable theory of sensitivity. Two reasons were offered in support of this claim. One was that theory of learning allows greater predictive power than the sensitization construct. To make this point he asserted what a theory of learning predicts about various experimental situations, raising parallel questions about what a theory of sensitivity would predict.[55] The other reason was superior explanatory power. The theoretical vacuity of sensitivity allows critics to mask and dismiss evidence that is best explained as true learning:

> If one wishes to explain all of the conditioning studies as showing mere sensitization due to noxious stimulation rather than showing true learning, then one must have in mind a type of sensitization that builds up slowly from one day to the next and that remains in rather full force for weeks or even months when the animal is not being trained, yet disappears in a matter of a few trials when the conditioned stimulus is presented without the unconditioned, and yet reappears very rapidly indeed when retraining is instituted. I ask you, what kind of "mere sensitization" is that? If one wishes to invoke sensitization to explain the Griffard and Peirce study (1964) or any of the other experiments showing conditioned discrimina-

> tion, then one must realize that one is advocating a type of sensitization that is clearly stimulus specific. Davis (1963) has shown that giving planarians electroconvulsive shock immediately after training knocks out the memory of the conditioning; yet would not one expect convulsive shock to increase rather than to decrease sensitization?[56]

McConnell's maneuvers in argument illustrate that in the logic of doing science, there exists the logic of what rhetoricians call burden of proof. A rhetorical-logical principle of managing burdens of proof where discussion concerns evidence is that questions about existence ("Is it?") and definition ("What is it?") have to be settled before it becomes reasonable to address qualitative issues. We need to know that something exists, and what it is, before we can relevantly argue about its qualities or the methods of deriving evidence about it. On the evidential exigence McConnell had tried to meet a burden of proof by (1) taking a *definitional* stand that evidence shows transfer of *learning*, and (2) urging that evidence at least warrants the *qualitative* judgment that *something* chemically transfers. Because the stoppage concerning definitions of evidential meaning still existed in the situation, he reframed the exigence as *interpretive* rather than evidential, and sought to shift the burden of proof back to the critics who rejected the memory-transfer hypothesis in favor of the sensitization hypothesis. Critics could not respond directly to McConnell's challenge without addressing (1) the definitional questions: What is the theoretical meaning of sensitization? and What is the theoretical meaning of learning? and (2) the qualitative issues: What experimental applications of learning are meaningful? and What experimental applications of sensitivity are meaningful? Only then could they address head-on the translative question McConnell was raising for them: Can sensitivity account more meaningfully for behavioral changes in conditioning studies than learning? That McConnell intended to create definitional and interpretive burdens for critics of memory-transfer research by framing the interpretive exigence as translative is clear from his summarizing ultimatum:

> In summary, I think it is about time that the theorists who hold to what Walker and Milton (1966) call "the sensitization hypothesis" either put up or shut up. They must tell us exactly what the hypothesis predicts in the various situations I have asked about, or they should admit that their sensitization hypothesis is fuzzy to the point of uselessness.[57]

McConnell placed a burden of proof on his critics that allowed him to shift the field of controversy from *evidential* problems to problems of *interpretation*. This shift would not very likely satisfy scientists who had *definitional* doubts about the meaning of evidence or *conjectural* doubts about the reliability of evidence. Nonetheless, the strategy does have some rhetorically logical plausibility. In a rhetorical situation beset by persistently conflicting interpretations of data there are two logical ways to dislodge the impasse: One can try to settle definitional questions about data either by (1) returning to conjectural issues concerning the reliability or relevance of the evidence, or (2) examining the meanings and applications of constructs used to interpret the evidence. It is clear some critics thought the best way to resolve the planarian controversy was to pursue the first option insofar as planarian conditioning studies were concerned. McConnell believed those contending learning and its transfer does not occur should focus on the second alternative to loosen argumentative blockages. However, McConnell did not believe this alternative would resolve scientifically valuable problems.

The blunt tone of McConnell's ultimatum was conciliated by: "Of course, anyone who thinks the problem through carefully will probably conclude that the battle is largely based on semantics."[58] He asserted that behavioral modifications are not likely to be due exclusively to learning, or sensitization, or habituation; but in fact the processes are so complex that all three constructs can be simultaneously involved: "It would be rather frightening if the *only* factors that transferred chemically were those due to the pairing of the conditioned and unconditioned stimuli or, on the other hand, if *none* of the factors that transfer were in any way influenced by the pairing of these stimuli."[59] McConnell was suggesting that planarian research had become so mired in nomenclature problems that scientists had lost sight of what was really scientifically significant—the facts concerning chemical transference of acquired bahavioral changes:

> For it seems to me that, when the theorists who hold to the sensitization hypothesis are forced to account for all the data, they will find they have swallowed the camel and are now straining at a worm. The fact that, in their terms, long-lasting sensitization occurs, that it is chemically mediated, that it shows many of the characteristics of so-called "true learning," and, most of all, that it can successfully be transferred from one organism to another via cannibalistic ingestion or RNA

> injections or both—these facts are so incredible that to argue whether this sensitization is or is not true learning seems to be emphasizing semantic trivialities rather than looking at data logically.[60]

McConnell was trying to shift discussion away from problems of interpretation back to what he considers to be fundamental facts. He did this in evaluative terms by resorting to the *topos* of fruitfulness. What was agreed upon (that *something* chemically transferred) could serve to produce new findings and insights. The phenomenon of transfer might yield detailed data about the chemical bases of behavior, which could, or at least should, encourage scientists to work at rearticulating or even inventing constructs more meaningful than the "semantic" ones creating confusion. McConnell's concluding paragraph was an explicit appeal to scientists to mend interpretive divisions and reunite on the shared value that it is comparatively more *fruitful* to pursue investigative work on the now-established chemical-transfer phenomenon than to divert attention to terminological wrangling:

> Let me then make a suggestion. If we stop arguing about whether worms can learn or whether their behavioral modifications are due to mere sensitization, we could start all over again and avoid bringing into comparative psychology all of the tedious, trivial arguments that have plagued experimental psychology since the time that Hull and Tolman took up their rats and began battling. We could begin by gathering correlations between environmental changes and behavioral changes in planarians, not bothering too much about what labels we assign to the processes involved. And, since we have an intriguing new tool, the transfer phenomenon, let us put it to work for us as cleverly as possible. Suppose we find, as I suspect will be the case, that some components of an acquired behavioral change will transfer, while others will not. Likewise, suppose we find that the various components of the acquired behavioral change are mediated by quite different chemicals. What an interesting lever such findings would give us in an attempt to pry out the details of what really goes on inside an organism when its behavior changes! Perhaps we shall eventually come up with new and better definitions of learning and sensitization. More likely, we may eventually coin new terms to cover our better understanding of the biological bases of behavior, terms that hopefully will lack the surplus meanings and emotional connotations of the words we have used too long in the

> past. I do believe that the current studies on the chemistry of memory could play a significant part in such an undertaking. It would be a real pity if we chose instead to continue the semantic battles and word play that so many of us have engaged in recently.[61]

In making this final appeal McConnell was once more engaging a qualitative *stasis,* but this time the point of stand concerned evaluative matters. It was certainly appropriate to address the question: Which would more fruitfully advance knowledge about the biochemical bases of behavior: continued investigation of the transfer phenomenon or continued discussion of interpretive problems? But in raising this question McConnell had already lost those who believed that planarians could not be trained reliably. Those people would have methodological objections and conjectural concerns about the reliability of evidence. Even those believing *something* transferred might see this claim as less fruitful than transfer of stimulus-specific *learning*—a phenomenon thought necessary to support the memory transfer hypothesis. These individuals would not be so willing to waive interpretive issues.

The biologist Patrick H. Wells agreed with McConnell's advice to focus on evaluative rather than interpretive issues, but he took a conjectural stand against the scientific significance of the transfer phenomenon insofar as it was established evidentially. Said Wells:

> I agree that it is difficult to differentiate operationally between the words learning and sensitization as they have been used by many writers. But, rather than argue about the meanings of words, we have devoted our attention to designing experiments which would test the validity of the hypothesis of memory molecules. Our main concern was that most of the transfer findings involved experiments in which light–shock pairings were used. Our reasoning was that, if there were an encoding of memories in a molecule, it ought to be possible to demonstrate the transfer of other kinds of learning.[62]

Wells granted that "in an early experiment we did obtain data to support the finding that, by allowing naive worms to cannibalize light–shock trained subjects, there was a more rapid conditioning of the naive worms."[63] However, said Wells, the transfer phenomenon, even if accepted, so lacked scope in application that it could not warrant a general hypothesis of a molecular memory code. He argued that transfer was shown only when worms trained in light–shock pairings were cannibalized; transfer did not occur when worms trained in T mazes or Van

Oye type mazes were cannibalized. If memories were encoded in molecules, one ought to be able to transfer these other kinds of learning. Wells concluded, "Based upon our findings, we would have to reject the general hypothesis of a molecular coding of memory."[64]

By arguing in this way Wells raised a conjectural *stasis* but with reference to evaluative considerations: the memory-transfer hypothesis was too narrowly supported to have scientific significance. He thus disagreed with McConnell's qualitative framing of the central evaluative issue.

Edward L. Bennett challenged the evidential grounds for McConnell's definitional claims that evidence can be interpreted as transfer of learning. In doing so he rejected McConnell's depiction of interpretive problems. Specifically, Bennett asserted that McConnell's experiments had been insufficiently replicated. He thus moved back to the most fundamental issue of whether there was reliable evidence for transfer of some form of learning. He claimed directly that McConnell's and related experiments were qualitatively poor:

> Dr. McConnell mentioned that he was not thinking very far ahead when he published his revolutionary results and that he should have hidden them under a barrel and made them out as something else. I would like to suggest that, if the key experiments had been replicated sufficiently and with a large enough population of subjects to begin with, we would all have been satisfied. The McConnell, Jacobson, and Kimble (1959) experiment used an n of about 5 in each group. The RNA transfer experiments that have been reported are poor, and, in the case of Fried and Horowitz (1964), the authors themselves admit this. If studies such as these look as if they might get us somewhere, why haven't they been exploited? Again, there are few repetitions. The Westerman (1963) experiment, e.g., is another key one that needs repeating. I don't know what has happened to Westerman, but surely he can see the potential of his finding. Why hasn't he exploited it?[65]

As this passage suggests, Bennett was making the crucial exigence one of methodology. If successful, he would bring debate back to its conjectural starting point: Is there or is there not reliable evidence to warrant the conclusions that worms can be trained and their memories can be chemically transferred? Bennett doubted it.

The analysis of *stases* in an extended scientific discussion illustrates what can be gained by exploring scientific argument using the *stasis* system that is peculiar to doing science. One important value is that the *stasis* system provides a heuristic for sorting out issues addressed in scientific discourse within quite complex rhetorical situations. Superior *stases* isolate four possible "master" exigences scientists can address and remain scientifically reasonable. Regardless of how exigences are framed, fundamental premises of *stasis* theory also state that in every rhetorical situation certain logical obstacles to agreement will be present, must be dealt with, and need to be dealt with in a logical sequence. Subordinate *stases* help in isolating those obstacles a rhetor thinks must be addressed to establish the situational reasonableness of his or her most important claims. The *stasis* system for science helped in tracking McConnell's framing and reframing of evidential, interpretive, and evaluative exigences. It also helped to identify the central logical obstacles McConnell hoped to dislodge or avert.

Analysis of *stasis* management shows that McConnell claimed his evidence and various corroborations showed (1) learning is stored chemically; (2) learning can be transferred chemically; (3) RNA is the likely transfer molecule; (4) even if (1)–(3) proved inadequate, the totality of evidence still warranted the claim that "something" chemically transfers; McConnell also claimed (5) learning accounted more effectively for planarian conditioning studies than sensitization, and (6) the now-established fact of chemical transfer is more significant scientifically than problems of interpretive nomenclature. These claims collectively indicate that McConnell thought the most situationally immovable obstacles blocking authorization of the memory-transfer phenomenon were *definitional* about evidential exigences and primarily *qualitative* and *translative* about interpretive ambiguities. Recognizing that he could not remove these obstacles from his rhetorical situation, he sought to conserve his fourth basic claim about the evidence sufficiently warranting the transfer phenomenon and his sixth evaluative claim that investigating the transfer phenomenon promises greater scientific fruitfulness than continual debates about definitional and interpretive problems. McConnell thus took his two strongest and most fundamental stands on qualitative issues about evidence and scientific significance.

We can discern from the ensuing discussion that some scientists disagreed with McConnell's framing of problems and related issues. Wells waived the definitional and interpretive issues, but made a *conjectural* challenge to McConnell's qualitative, evaluative claim. It is clear Wells thought McConnell's qualitative appeal was logically premature. Bennett thought all discussion of evidential meaning and sufficiency was logically premature because *methodological* problems made the reliability of supporting evidence doubtful. By suggesting the conjectural issue about evidence had not yet been settled, he could imply McConnell's definitional and qualitative claims about evidence were situationally unreasonable and logically premature.

McConnell's stands and the points of clash raised in subsequent discussion allow us to illustrate how the logic of *stasis* analysis can be used prescriptively to remove communicative blockages. The point of *stasis* theory in general rhetoric is that by analyzing the probable points of *stasis* one discovers both which issues must be met, and in what order, to further reasonable discussion. I contend the same principle applies in the rhetoric of science. To resolve the evaluative impasse with Wells, both he and McConnell would need to work through conjectural, definitional, qualitative, and perhaps translative obstacles regarding scientific value and do so in that order. Any hope of removing the communicative blockage with Bennett would require that both scientists settle definitional and qualitative questions concerning experimental standards and procedures. There can be no genuine progress in either of those clashes unless arguments are systematically focused on the same issues and in a situationally logical order so that resolving agreements can be reached. Otherwise scientists could argue indefinitely at cross purposes without ever achieving a genuine, resolvable point of *stasis*.

Examining the *stases* that rhetors address can help critics of scientific rhetoric to estimate the wisdom of rhetorical choices. The system of *stasis* analysis for scientific rhetoric not only allows (1) understanding the strategic choices rhetors did make, but also (2) consideration of alternative possible choices they could have made. The critical contrast reveals options that might have enhanced the situational reasonableness of claims actually made. Suppose that instead of making all six claims McConnell had focused attention on the thesis that transfer of something occurs in most experiments and that this fact was so significant it deserved attention. To accomplish this thesis he might usefully focus on conjectural issues about evidence, interpretation, and evaluation. He could thus claim (1) chemical transfer of something occurs, (2) terms like "learning" or "sensitivity" are too imprecise for full understanding of this phenomenon, and (3) the transfer phenomenon itself can serve fruitfully as a lever for working out new constructs grounded in chemically based facts. This plan for *stasis* management would (1) reduce the range of potential disagreement by keeping discussion focused on evidential grounds, and (2) make a more logically convincing request to waive controversial interpretive and definitional issues by keeping the points at issue conjectural. Or suppose that McConnell focused his entire presentation on the interpretive problem of what "learning" means. He might address conjectural, definitional, and qualitative issues about this interpretive exigence by claiming (1) present psychological theories of learning inadequately account for the transfer phenomenon; (2) a chemically based theory of learning can be formulated to fill the resulting interpretive void; and (3) this theory can emerge fully only through careful investigation of the transfer phenomenon in a variety of experimental settings (in which correlations are sought between different stimuli and chemical changes and between chemical changes and behavioral changes). Again, the presentation could have held subsequent discussion to the evidential grounds that ordinarily constitute the first grounds of *stasis* in scientific discussion. Though I cannot assess McConnell's motives for choosing the strategies that he did, I can suggest that given the logic of *stasis* theory, the effort to address six different issues enhanced the prospect that subsequent discussion would be multifaceted and the most fundamental issues about the existence of relevant evidence, its meaning, and its significance would remain unsettled.

Examining the *stases* that rhetors address can also help an observer decide whether they responded fittingly to situational exigences. When, for example, scientific questions are framed as conjectural with reference to evidence, a knowledgeable critic can anticipate the character of fitting rhetoric in response. One fitting response to such a question would be to show that experimental replications have yielded corroborating data. Relative to his regeneration, cannibalism, and RNA studies McConnell did this, and we should expect that this kind of response would be judged reasonable since he was trying to resolve conjectural matters relevant to available evidence. We could also observe that qualitative and translative issues about interpretive constructs are fittingly engaged with comparative assessments of predictive and explanatory power. Fruitfulness, too, is an appropriate evaluative theme for engaging issues about the significance of claims. By contrast, however, McConnell's ad hominem attacks on critics were surely not the most fitting responses to conjectural or definitional issues about evidence. A scientist can scarcely settle the question of whether there is evidence, say, of transfer of learning by insinuating that those who question the evidence or its meaning are pigheaded. Unless challenges to scientific credibility are supported by relevant arguments and evidence, attacks on the ethos of other scientists will seem at best peripheral to genuine points at issue and at worst extrinsic to the tacit but established guidelines for reasonable scientific discourse.

When an observer examines the points of *stasis* addressed, it is further possible to determine whether the point of *stasis* is a point of genuine disagreement or one of semantic misunderstanding. Genuine points of *stasis* in science can be resolved only when rhetors supply arguments and evidence sufficient to persuade other scientists that one position or another ought to be adhered to. In analyzing such practices we are not observing the kind of dialectic in which the object is to force an opponent to concede or leave; we are observing discourse in which the objective is to earn endorsement from a particular community of persons committed to maintaining and expanding their knowledge. In respect to McConnell and his critics we can now see that McConnell's argument and evidence for chemical transference of learned/conditioned behavior constituted the weakest aspect of his contention that chemical transfer took place. And this was, of course, an issue that remained open at the end of the arguments. By searching out how *stasis* issues were managed, we can see why that point of *stasis* remained unresolved.

Stasis analysis also helps to distinguish alternative orientations that scientists might take toward evidence, constructs, values, and methods. In this rhetorical situation there was no consensus on the meaning of "learning".[66] Observers and participants therefore could not assume the term would be used in the same way by all discussants. Indeed, the term could shift in meaning within a single rhetor's discourse. This is illustrated by McConnell's argumentation. McConnell asserted the controversy over learning versus sensitization focused on a semantic, not a substantive, issue. This seems odd given that he had treated the issue as serious and genuine earlier in his discourse. However, when he shifted from interpretive issues about learning and sensitivity to his evaluative common-ground stand about the fruitfulness of the transfer phenomenon, he was switching orientations for understanding learning: the semantic issue concerned learning as it is understood within the theoretical orientation of stimulus-response psychology; the substantive issue was how to forge a chemical theory of learning based on the transfer phenomenon.[67] McConnell's use of apparently neutral expressions like "acquired behavioral tendencies" can therefore be interpreted meaningfully from the vantage of at least two different theoretical orientations toward "learning." To resolve definitional and interpretive issues firmly it is thus necessary to distinguish the theoretical orientations at work in the rhetorical situation and determine how each should (or should not) be understood and applied in planarian memory-transfer research. *Stasis* analysis provides a heuristic method for identifying and sorting out these and other underlying differences among discussants.

Stasis analysis helps underscore the point that using scientific methods involves more than following a set of unambiguous procedural instructions. Collins made the important claim that experimental replication requires "enculturation" into the often tacit standards for assessing what counts as a working experiment, rather than just fidelity in following some neutral procedural algorithm.[68] Travis extended this claim by pointing to procedural ambiguities as central in fueling the planarian memory-transfer controversy.[69] From a rhetorical perspective, persuasive applications of methods involve social negotiations about what can and what cannot be reasonably expected within problematic rhetorical situations. *Stasis* analysis indicates the range of potential methodological points for such negotiation.

McConnell directly addressed methodological issues in a second paper delivered at the symposium. Analysis of this presentation and subsequent discussion provides clear examples of disagreement about precisely which methodological points needed adjudication for scientifically reasonable research progress. McConnell's paper was titled "Specific Factors Influencing Planarian Behavior."[70] The paper held that relevant variables influencing planarian behavior are not accounted for in much of the planarian research. He opened the paper by saying:

> Let me begin by stating what I believe to be a terribly important rule of experimental psychology: In at least 90% of the studies published in psychological journals, the most important factors influencing the animals' behavior were those factors the investigators did not measure, did not control, did not mention in their written reports, and probably did not even know about. The rest of the paper is an attempt to prove the validity of this particular point with specific reference to the planarian literature.[71]

McConnell reviewed numerous factors that actually or potentially influence the behavior of planarians and claimed that experiments failing to produce evidence of conditioned behavior were conducted with insufficient attention to replicating controls over these variables. They should, then, not be counted as replications. For instance, McConnell claimed worms prefer slimed to clean surfaces, and that failure to control for this variable explains why many scientists were unable to condition their worms. He concluded, "All results obtained from animals trained in clean apparatus may well have yielded spurious results."[72] The ensuing discussion made it plain that some in McConnell's audience felt that problems of trying to replicate experiments were due to definitional inadequacies in formulations of experimental specifications rather than to procedural quality of their execution. Some said it was up to McConnell and his colleagues, as scientists, to make their work replicable through offering clear definitions.

E. Halas, a psychologist from the University of North Dakota, explicitly shifted the issue from quality to definition with a good deal of force:

> McConnell has stated that 90% of the articles in scientific literature are difficult to replicate, and therefore it is not too surprising that his studies are difficult to replicate. I don't know where or how he obtains a percentage of 90%, but I agree that at least some experiments are difficult to repeat. But like it or not, as far as science is concerned, replication is "the name of the game," and I would not want to see the day when we accept data on the basis of faith rather than objective replication.[73]

Halas struck directly at what he perceived as the definitional inadequacy of McConnell's and his associates' work:

> I was very interested in Dr. Corning's comments about procedure and that my procedures were not like the McConnell and Jacobson methods. Which procedures should I follow? They change their procedures every month, and they are still changing. If they had successful methods back in 1955 and 1959, why didn't they keep them?[74]

At this point Corning, a psychologist from Fordham University, defended the quality of McConnell's procedures, saying: "I think the reply would have to be that, rather than beat a dead horse like the Thompson and McConnell paper, it would be better to improve methods, as McConnell has done and we have done, and get results which was our experience."[75]

Clearly, what was being wrestled with was whether the problem of methods was qualitative or definitional. One immediate point to notice is that if McConnell and his supporters kept the issue qualitative, discussion of comparative qualities could continue indefinitely; but if the opponents could force definition to the center of discussion, a clear-cut stand would occur until the McConnell group established their own clarity to the satisfaction of the critics. This seems exactly what Bennett tried to accomplish when defending his (and Calvin's) highly influential claim that planarians could not be trained reliably:

> The thing we were trying to emphasize is that there are many factors which are involved in planarian training, if indeed they can be trained. It would be useful if these factors could be consistently described, especially for the biochemist who doesn't want to spend his life training. Some variables are known, but there are some inconsistencies. Jacobson reports a 60% level of response with training, yet, in studies in which people have not been able to repeat the regeneration and cannibalism findings, we see the criticism that they had not trained their animals to a high enough level. At first we were told to look for contractions, turnings, and what-have-you. Now, we are told that only contractions can be considered. Dr. Corning tells us that mass conditioning is fine, yet McConnell criticizes negative studies on the basis that they have used mass conditioning. In some reports we see that light increases responses; in others we see that there is habituation. We are told by some to train in the morning, others advocate the afternoon, and still another tells us to train according to whether the moon is out. I will not deny that any of these factors are important, but how many of these reports represent experiments that have been replicated several times? We need evidence of interexperiment reliability. And one wonders why there are all these inconsistencies.[76]

When scientists engage in debate about methodological problems in a rhetorical situation involving evaluation of quite controversial and potentially revolutionary claims, we can expect that the rhetorical logic of burden of proof will favor scientific skeptics. McConnell and his supporters did address definitional issues in their published work, but they could not dislodge these obstacles sufficiently to take fully convincing qualitative stands in this rhetorical situation. Given the problems of replication, efforts to shift discussion back to definitional issues were likely to seem situationally reasonable even to some proponents of planarian learning.[77] In doing science precision and clarity about procedures and the reasons for them are values that arise before one gets to the qualities of their execution.

CONCLUSION

All rhetors, including scientists, need to address exigencies that their situated audiences recognize as important. Obstacles or ambiguities constitute exigencies peculiar to scientific discussion because they impede efforts to maintain or expand a scientific community's comprehension of natural order. Scientific obstacles or ambiguities appear to be of four sorts: evidential, interpretive, evaluative, and methodological. I have called these classes of difficulties the superior *stases* of scientific rhetoric. Within each of these classes of difficulty, specific kinds of intellectual obstacles occur and, individually or in combination, constitute the conjectural, definitional, qualitative, and translative points for judgment in any rhetorical situation involving the doing of science.

These points for judgment can occur in fairly predictable ways in recurring rhetorical situations that confront scientists. For example, when nonconventional or revolutionary claims are offered, we should expect defenders of orthodoxy to make their strongest stands on *conjectural* issues respecting *evidential, interpretive, evaluative,* and *methodological* matters. If defenders can preserve orthodoxy on all four of these points of *stasis,* a novel claim can hardly seem scientifically reasonable. Consider the options if someone claims to have discovered a marine animal called a sea serpent. The claim can be challenged on grounds that there is no hard evidence such as a clear photograph or the remains of any such animal. The challenge would then be conjectural relative to scientific *evidence.* The meaningfulness of the claim could be challenged on grounds that taxonomic classifications give no place to such an animal. The challenge would still be conjectural but relative to scientific *interpretation.* Were someone to claim the animal had a serpentine neck and a dorsal fin resembling that of a shark, a reply could be that such an animal is so anomalous that the sighting lacks scientific significance. The issue would again be conjectural but now relative to *evaluation.* The general sea serpent claim can be attacked on grounds that the sightings were not obtained through systematic, expert observation, and so are not genuinely scientific. Once more, the issue would be conjectural but now in reference to *methodology.* These are the specific kinds of rhetorical problems all radical scientific claim-makers face.[78] The obstacles to radicalism in science hinge specially on conjectural issues concerning evidence, interpretation, evaluation, and method. Examining how scientists respond to other kinds of situational exigences should reveal additional standardized strategies of *stasis* management.

Analysis of *stases* focuses attention on central points of difference regarding existence, meanings, significance, and actions. Once such points are isolated, scientists can at least understand each other's claims, whether they agree or not. This sort of analysis can further communication among adherents of quite different paradigms and orientations, allowing them to "make sense" of each other's claims. *Stasis* analysis can accomplish this because the scientific *stasis* categories identify the kinds of judgments scientists make without presupposing the specific contents of those judgments. *Stasis* categories provide substantively neutral commonplaces that make it possible for those adhering to different orientations to express their stands with minimal bias and distortion. They thus help discussants to trace sources of communicative conflict or confusion by investigating how responses to the sixteen fundamental *stases* converge and diverge. Doing this makes disagreement among those embracing different orientations logical insofar as the *stasis* categories help discussants discover what disagreement is about. Collectively the scientific *stases* constitute a rhetorical procedure for working out understandings of where discussants substantively agree and disagree about problems of empirical existence, scientific meaning, scientific value, and scientific action. *Stasis* analysis could very well be the method for resolving communication problems arising among scientists who embrace alternative paradigms and other orientations.

Sometimes it is charged that *stasis* theory in rhetoric only yields an array of technical classifications whose usefulness in practical affairs is at best questionable and at worst destructive of inventional creativity.[79] I have shown these charges are untrue in a variety of ways. I conclude with the further claim that commentators on the history and sociology of science inevitably, perhaps unconsciously, apply parts of *stasis* theory.

Clashes in scientific thought are primary grist for sociologists and historians of science. Such scholars must analyze the stands, or *stases,* on which arguments and counterarguments focus. It is not surprising they do not put to use the comprehensive theory of rhetorical stands. *Stasis* theory was lost to Western culture with the Age of Enlightenment and, especially, the Age of Romanticism. Nonetheless, if one examines Travis's studies of the memory-transfer controversy, Collins's analysis of Joseph Weber's claims to have found high fluxes of gravity waves, and Collins's and Pinch's investigation of arguments for and against the scientific legitimacy of parapsychology, one discovers that the commentators explain what happened by pointing to the primary points of clash between disputants. *Stasis* analysis can extend these and other analyses by providing a systematic method for investigating scientific discursive controversies. Using the *stasis* framework heuristically can help commentators (1) distinguish the kinds of stoppage that fueled controversy, (2) explore whether the controversialists made use of all available persuasive resources to sustain their argumentative stands on those issues, and (3) isolate the precise *stasis* questions that had to be answered if scientific discussion was to advance. In science, as in other domains, wrong turns were made and right turns were delayed because those engaged in the rhetoric of science focused on noncrucial points of stoppage, failed to present their claims in strongest forms, and spoke or wrote beside the points that really needed adjudication. Bringing to bear the full theory of *stases* outlined in this book can help historians and sociologists and rhetoricians of science fill in such gaps in understanding. Doing so can also equip scientists to carry on scientific discourse more knowingly and more incisively.

To a major degree "doing science" is social construction and destruction. It is social action aimed at inducing situated audiences to authorize what is said as scientifically reasonable. To accomplish this, scientists have to decide how best to respond to exigences that arise in scientific work. This means picking out the crucial points for decision, expressing them so they secure needed attention, and rendering scientifically reasonable the decisions proposed to the community. Those processes are rhetorical, but they are not illogical. Skillful, effective execution of them begins with scientists' locating and understanding the issues, and then discovering available, relevant, persuasive lines of thought—lines characterized by the *topoi* of scientific discourse. The role of topical choices in creating influential scientific communication is the subject of the next chapter.

9

RHETORICAL INVENTION IN SCIENTIFIC DISCOURSE: Discovering Lines of Argument

As I have already indicated, *topoi* are headings or topics that identify lines of thought. The headings specify alternative ways of treating ideas. Some of these lines of development are relevant to virtually all discussion within a culture; others are peculiar to specific subjects and fields of inquiry. Once one has discovered what significant points of *stasis* need to be addressed, the next question is, What is the best way of addressing these exigences in the situation in which the communication will be heard or read? The general *topoi* used in one's culture and the special *topoi* peculiar to one's subject and field of inquiry suggest what communicative possibilities exist in a given rhetorical situation. This is as true in scientific communication as it is in other rhetorical enterprises.

There are topics of discussion that characterize doing science regardless of specialty. This is where I shall focus attention. I shall not deal in any detail with *topoi* peculiar to particular sciences because that would require technological knowledge of each science and its methods. However, each of the scientific *topoi* I shall identify can be reformulated with greater technical precision to identify lines of thought peculiar to any specific science.

Scientific rhetors do in fact draw formulations of propositions from experience and from received knowledge, and they select and use those formulations likely to yield the best possibilities of rhetorical success within situational constraints confronting them. Scientists may not be conscious that they do this, but they do it nonetheless. Two consequences flow from this: One can extract the rhetorical *topoi* of scientific discussion from scientific discourse, and scientists can become more astute and inventive in carrying on their discussions if they know and consciously canvass the standard *topoi* of scientific discussion.

I propose to identify and discuss the character of the *topoi* that inform scientific rhetoric. In the process I shall consider whether the *topoi* are more scientifically field dependent, and therefore special *topoi,* or are more field invariant, and hence general *topoi.*

PROBLEM-SOLUTION *TOPOI*

Lines of thought that relate to solving problems inevitably arise in scientific discussions. I shall call these problem-solution *topoi.* These *topoi* reflect prominent features of the special, field-dependent, problem-solving logic governing scientific inquiry and discussion. They thus constitute ways of talking that treat claims positively or negatively on the basis of whether the claims help to solve scientific problems in scientifically reasonable ways. Generally speaking, the problem-solution lines of thought in science are used to establish (or disestablish) firm connections between observational or theoretical claims and what is accounted for by accumulated data or theory. The arguments of this class appeal to scientific communities' interests in solving problems within the frameworks of agreed-upon observations and theories and in accordance with scientific methods. Accordingly, a scientific rhetor using problem-solution arguments must know what the community seeks to solve and what it agrees upon as legitimate observations, theories and methods. Claims need to be consistent with both. A number of subclasses of this kind of argumentation occur in scientific discourse.

Experimental competence is an especially important problem-solution *topos.* This *topos* suggests lines of argument that attack or defend data and claims on the grounds of perceived competence of experimenters as well as the quality of experiments.

In chapter 7 I pointed to Sebeok's attack on Patterson's research and to arguments leveled against the claims of parapsychologists. Many of those specific arguments derived their persuasive force from whether research done was done competently enough to help to solve the scientific communities' problems. Such a theme is not used only against those who seek to change scientific judgments in major ways. Everyone who reads any amount of scientific literature encounters numerous discussions of the tightness of experimental designs, of whether control groups were sealed off from experimental influences, or the like. These are all specific developments of ideas about the problem-solving strengths of experimental competence.

An illustration of the use of experimental competence as a persuasive *topos* is provided in Harvey's study of experimental tests of quantum mechanics.[1] Holt, a PhD student, performed an experiment whose results contradicted predictions of quantum mechanics and lent support to the more peripheral and much less popular "local hidden variable" theory. This surprising result created a difficult rhetorical situation for Holt. Overwhelming presumption favored quantum mechanics. What could he do, since his claim against the predictive potency of quantum mechanics would be unlikely to be accepted as reasonable by most scientists? He could not easily discuss his result as an error because that would diminish perceptions of his experimental competence—not a happy prospect for an experimental physicist fresh from graduate school and looking for employment. He could simply present the results and move on to some other project, but this option would be unlikely to convince interested scientists that a PhD candidate could find a problem in quantum mechanics other researchers could not. (There was, in fact, a similar experiment that contradicted Holt's findings.) He could try to publish his results as findings that accurately indicated the presence of a significant anomaly in quantum mechanics. The centrality of quantum mechanics in physics and the marginality of local hidden variable theory made this alternative undesirable, especially since corroboration of his results would be slow in coming if it came at all. If he suggested that his results could be due to faults in his experimental apparatus, he would certainly have difficulties in making requisite explanations to his PhD committee. Holt's rhetorical situation was rendered still more complex by the facts that he did not believe the results himself and that his PhD supervisor once had a student who obtained and published results conflicting with accepted concepts in quantum electrodynamics, only to have the results turn out to be spurious.[2]

Holt chose to argue that his results were due to a persistent error in his experimental apparatus. He did not hope to render his findings reasonable through experimental corroboration, nor did he ask acceptance of those results as validly indicating a significant anomaly. He opted for a course of argument that yielded the best possibilities for inducing scientists to perceive him as a competent experimenter. Holt spent two years attempting to isolate the source of error in his apparatus. He did so to save his reputation and his doctorate by showing in his words and deeds that he did experimentally competent research. In Holt's words, "I had to spend an extra two years looking for systematic errors to make sure that anybody would believe me."[3] Holt was successful in preserving his rhetorical ethos, and he thereby preserved perceptions of his value as an experimentally competent problem-solver, as is indicated by Harvey's assessment of the outcome of Holt's choices in this rhetorical situation:

> Holt's actual response was clearly a sensible one. Faced with a result which he did not believe, his actions seemed to him to be a way of minimizing any damage which could be done, and of optimising the outcome of what could have been a very embarrassing episode. By showing that the error was persistent and apparently deeply-rooted in the apparatus, and by permitting other physicists to examine his apparatus, without isolating the error, he attempted, as it were, to "defuse" the error so that it did not reflect seriously on his own competence as an experimenter. My impression from talking to nearly all of the LHV [local hidden variable] physicists is that Holt has successfully presented himself as a good experimenter who had a bit of bad luck, obtained an incorrect result, yet treated that result in the correct skeptical manner.[4]

Holt's predicament, the course of self-presentation that he chose, and the approval of fellow scientists all indicate how critical arguments (and deeds) relating to the *topos* of experimental competence are in the conduct of science. Empirical orientation in any science renders this theme or the related theme of observational competence inescapable in scientific discourse.

Observational competence is an important problem-solution *topos,* especially in sciences that cannot make and collect data through experimental intervention and control. Freeman's recent attacks on Margaret Mead's early work with adolescent Samoan girls constitute, in part, a challenge to that then-youthful anthropologist's observational competence.[5] Freeman claims Mead lacked training and experience with field methods. He alleges that she was fooled by her young interviewees because their responses were more likely attempts to please Mead by telling her what the direction of her questions implied she wanted to hear. Further, Freeman points out that Mead had a preexisting commitment to the cultural determinism of her mentor, Franz Boas. By documenting an easygoing adolescence quite unlike that of American youth, Mead's *Coming of Age in Samoa* supplied powerful evidence for cultural determinism and against genetic determinism. Freeman uses these reasons, among others, to place Mead's case for cultural determinism on questionable empirical grounds. The reasons persuade insofar as they create ambiguities regarding Mead's competence as an observer of Samoan culture.

Another familiar problem-solution *topos* is *experimental replication.* This *topos* should not be confused with the *topos* of corroboration or confirmation, which I will discuss shortly. Collins points out the logical requirements for using this *topos:* "When a scientist claims that an experiment has been properly replicated he is claiming that these two sets of events, the original and the replication, should be treated as the *same.*"[6]

Scientists argue for or against the reasonableness of experimental data by assessing whether it is possible to repeat an experiment or whether experimental decisions involving the design and implementation of experiments are sufficiently similar to warrant comparison of the results. Scientific rhetors are expected to provide sufficiently detailed explanations of methods and procedures so that the experiment can be replicated accurately.

As an illustration of how experimental replication can be used as a topic for attacking experimental results, consider Alexander's reason for rejecting Nobel Laureate C. G. Barkla's claim concerning a set of atomic radiations known as the J phenomenon:

> It was thought desirable to repeat as far as possible the experiments of Barkla and others, with the view of obtaining further information on the J phenomenon, but on account of the fact that in none of the papers published in support of the effect has there been given a *full description of experimental arrangements,* it has been a matter of some difficulty to ensure repetitions of previous work.[7]

The question becomes, Has Barkla really established or solved any problem?

When one considers the number of article citations scientific rhetors give to corroborate or disconfirm claims, it becomes very clear that scientists actually attempt to replicate few of the experiments they cite. It is in fact never possible to reconstruct exactly the decisions that were made while conducting an experiment. This is true regardless of how good the published specifications are. This difficulty gives rise to a different line of argument on the theme of replicability. Rhetors assess whether any two experiments are sufficiently similar to count as confirming one another.

I have mentioned in chapter 8 Weber's claim about having detected high fluxes of gravity waves. He argued that negative data used to refute his claim were produced by experiments that did not genuinely replicate his own.[8] In this usage the *topos* of experimental replication suggests reasons for rejecting data as not pertinent. On the other hand, evidence used against Holt's surprising results was thought reasonable in part because it was assumed that the experiment yielding those conflicting results was an adequate repetition of Holt's. Many scientists concluded straightforwardly that Holt's experiment had been repeated by Clauser. Even Holt accepted Clauser's experiment as refuting his own results. He assumed that there had been sufficient replication:

> Now that Clauser has done the experiment with the same exact cascade in mercury, with the same sort of lamp, I have to believe that it's got to have been the apparatus which caused the result. *It couldn't have been the physics because everything is the same there.*[9]

In fact, Clauser's published report carefully delineated both similarities and differences between the two experiments. Interested scientists evidently found it more reasonable to stress the similarities than the differences.

Another problem-solution *topos* that supplies good reasons for scientific claims is *experimental originality*. Even if it were possible to replicate exactly, it can be argued that to do so would be less than desirable. It is widely thought that good experiments should make modifications in the procedures of previous experiments in order to elicit additional information about variables that can influence experimental outcomes. Whether and when this is true are themselves matters for argument. For instance, in chapter 8 we examined an exchange between Halas and Corning about methodological exigences in planarian research. After Halas argued that continual changes in experimental procedures indicated replicability problems, Corning defended those changes as improving methods and achieving results. The basis of Corning's defense was that originality was required to bring ongoing experimental problems to solution. Claims of replicability and claims of originality are not, however, incompatible, as the Holt-Clauser case illustrates. Clauser's replication contained differences from Holt's practice, and those differences, in principle, could have constituted grounds for arguing that Clauser's experiment did not sufficiently replicate Holt's.[10] This did not occur because nontrivial differences were thought likely to elicit information about the experimental source of Holt's probable error. In this case experimental differences were thought reasonable because interested scientists were satisfied that the problem deserving attention was the source of Holt's surprising results, and not, say, potential anomaly in quantum mechanics. If Clauser had used exactly the *same* procedures as Holt, his experiment could not help to reveal the source of the presupposed error. The rhetorical principle operating in the Halas-Corning and Holt-Clauser cases is that experimental modifications and deviations are more likely to be adjudged reasonable when they are thought to yield insight into the problems a community seeks to solve.

Journal editors want to publish experimental research that makes original contributions to their fields. They therefore take umbrage at what they call LPU (Least Publishable Unit) articles.[11] As Broad explained, "LPU is a euphemism in some circles for the fragmentation of data. A researcher publishes four short papers rather than one long one." Evelyn S. Meyers, as managing editor of the *American Journal of Psychiatry*, said good reviewers should warn editors when a manuscript "is only a little bit different from what's been published." Thus experimental originality is (or in editors' opinions ought to be) a central theme for adjudicating what experimental research is worth publishing.

Corroboration is an especially important problem-solution *topos*. Corroborative arguments are among the most compelling, for if scientific claims are to stick, confirmations must be supplied at some point during their development. The *topos* of corroboration suggests at least two lines of thought subject to numerous specific adaptations. The first leads to acceptance or rejection of data claims on the grounds that they have or have not been confirmed by the results of other experiments or observations. In chapter 8 we saw how McConnell tried to strengthen the credibility of his evidential claims by citing experiments allegedly yielding corroborating data. Scientists used Clauser's experimental results as a good reason for rejecting Holt's findings. Holt was not corroborated. The second line of thought emphasizes that data support or refute hypotheses consonant with important theoretical constructs. Clauser's experimental results were accepted as additional confirmation for quantum theory and as refutation of local hidden variable theory. McConnell claimed evidence from planarian conditioning experiments tended to support the memory-transfer hypothesis rather than the sensitization hypothesis.

Arguments from the *topoi* of experimental competence, observational competence, experimental replication, experimental originality, and corroboration are potentially logical, structural choices whenever a scientist wants to establish or undermine the empirical bases of claims. Although applicable to all claim making they are especially useful in testing the credibility of data claims. A different way of arguing about the credibility of claims is to consider their scientific usefulness and theoretical fertility in accounting for empirical phenomena. Then we are likely to see arguments based on explanatory power, predictive power, and taxonomic power.

Explanatory power is a widely used problem-solution line of argument about whether claims supply and/or are consistent with propositions that tell us why certain specifics are observed. An argument of this kind needs to show that what is being explained *had* to be as reported, in view of broad and accepted propositions. The broader and more unqualified the theoretical proposition(s), the stronger this type of argument will be. Hence, alleging that a reported phenomenon is inconsistent with accepted principles of gravitational pull can be a powerful refutational argument. On the other hand, alleging a phenomenon is inconsistent with Malthusian hypotheses would be weak because Malthusian propositions are no longer generally accepted as scientifically true.

As an example of how claims can be established on grounds that they have explanatory power, consider the announcement by Hewish et al. of the discovery of pulsating radio sources, now known as pulsars. During an investigation of scintillating radio sources with a sensitive radio telescope a research student, Jocelyn Bell, claimed to have found on a recorder chart "some scruff which I found difficult to classify."[12] Hewish directed further investigation of the pulsing signals, whose discrete occurrences (there were gaps during which no pulses occurred) were apparently not associated with any known type of astrophysical object. Pulses also were spaced at intervals of about every one and one-third second, a measurement which, according to Bell's testimony, appeared to be "a very much man-made-interval, a suspiciously round figure."[13] Scientists were left with possibilities that the sources of the signals were terrestrial or that they resulted from man-made transmissions. Both of these explanatory alternatives were soon discounted after further investigations favored the conclusion that the pulses were a natural phenomenon of extraterrestrial origin. Accordingly, in the *Nature* article announcing the discovery, the authors first claimed they possessed data showing that the source "could not be terresterial in origin."[14] After a paragraph describing the unusual interval of the pulses, the authors continued by rejecting alternative interpretations and by tentatively stating a view alleged to have comparatively better explanatory power:

> The remarkable nature of these signals at first suggested an origin in terms of man-made transmissions which might arise from deep space probes, planetary radar or the reflexion of terrestrial signals from the Moon. None of these interpreta-

> tions can, however, be accepted because the absence of any parallax shows that the source lies far outside the solar system. A preliminary search for further pulsating sources has already revealed the presence of three others having remarkably similar properties which suggests that this type of source may be relatively common at a low flux density. A tentative explanation of these unusual sources in terms of the stable oscillations of white dwarf or neutron stars is proposed.[15]

In short, the authors argued that their tentative conclusion ought to be favored because, as theory, it explained better than any alternative theory.

The problem-solution *topos* of *predictive power* is drawn upon when theoretical propositions are attacked or defended on the basis of their capacity to capture empirical regularities that indicate what is going to happen. Claims have predictive power if they can orient scientists to anticipate phenomenal behavior not yet observed. When it can be shown that predicted relationships have been empirically confirmed, that argument will supply a powerful resource for rhetorical inducement. Ziman describes the power of this line of thought:

> What we say to ourselves is: if he did not already know the answer to the hypothetical experiment, then he could not have "cooked" his theory to agree with it; therefore it is much more convincing. The element of predictability in good science is just what makes us believe in it so strongly.[16]

The strongest argument for undertaking a specific program of research is that existing knowledge predicts the program is likely to lead to a discovery. For example, the efforts to "discover" the neutrino, radio waves, and the elements which filled empty spaces in the periodic table were largely justified as scientifically reasonable because the existence of all these objects was predicted beforehand by agreed-upon theoretical frameworks.[17] Conversely, unexpected discoveries that occur as violations of theoretical expectations, yet cannot be easily challenged, make the predictive power of the theoretical concepts suspect. Rethinking becomes necessary. For instance, discoveries of oxygen, Uranus, and x-rays were all unanticipated by shared theory, and adjustments had to be made in theory before strong predictive arguments could again be developed. As Kuhn put it, these discoveries developed from anomalies in theoretical expectations. Once the discoveries were confirmed and theoretical adjustments made, the adjustments provided "a new view of some previously familiar objects and simultaneously . . . [changed] the way in which even some traditional parts of science are practiced."[18]

It is fundamental to scientific thinking and argument that prediction be the acid test of claims about what it is reasonable to think or do. If proposals and claims do not ultimately promise clearer or firmer predictions, they are virtually irrelevant (at least as science is conducted in the West).

When interparadigmatic disputes arise, predictive power becomes a central issue. Adherence to claims for a new paradigm can be induced if proponents of the paradigm can show that their concepts predict phenomena or interpretations wholly unsuspected if one works within the traditional paradigm. Of course, making such claims and predictions credibly may not be easy. Credibility may come only after events create a new situation in which the claims of prediction can be more clearly supported. The credibility of Copernican theory was a case in point:

> Copernicus' theory . . . suggested that planets should be like the earth, that Venus should show phases, and that the universe must be vastly larger than had previously been supposed. As a result, when sixty years after his death the telescope suddenly displayed mountains on the moon, the phases of Venus, and an immense number of previously unsuspected stars, those observations brought the new theory a great many converts, particularly among non-astronomers.[19]

Taxonomic power is a problem-solution *topos* that suggests arguments for contending that experimental or observational claims can or cannot be identified and meaningfully categorized. Taxonomies have persuasive power when their categories are assumed to be individually discrete and collectively exhaustive. On these grounds taxonomic arguments are made to establish or to subvert the scientific reasonableness of claims. Nature studies provide numerous examples. For instance, naturalists are not interested in determining whether people claiming to have sighted sea serpents have actually discovered some large marine creatures. One reason is that descriptions of the animals vary so much that the creatures could not belong to the same species. As described, the beasts do not belong to a class. One individual believes that this problem of interpretation can be resolved with the claim that there are seven different kinds of sea serpent. That argument too rests on the notion that the taxonomical problem is what stands in the way of belief. Turning the argument around, others could dismiss data supporting the existence of sea serpents by contending that, at best, reports of sea serpents are misperceptions of animals already known and classified.[20]

Quantitative precision is a powerful problem-solution *topos* that can be used to strengthen theoretical or evidential claims. Kuhn identified this type of argument as decisive in inducing support for major scientific theories:

> The quantitative superiority of Kepler's Rudolphine tables to all those computed from the Ptolemaic theory was a major factor in the conversion of astronomers to Copernicanism. Newton's success in predicting quantitative astronomical observations was probably the single most important reason for his theory's triumph over its more reasonable but uniformly qualitative competitors. . . . The striking quantitative success of both Planck's radiation law and the Bohr atom quickly persuaded many physicists to adopt them even though . . . both these contributions created many more problems than they solved.[21]

The value of quantitative precision also is called upon to establish the credibility of data claims. After the announcement of the discovery of the pulsating sources, for example, scientists focused their attention on getting data about them with greater quantitative precision. Scientists at Jodrell Bank believed they could secure more precise measurements with their telescope than could the Cambridge group who had announced the discovery. Bernard Lovell's account of this period of research and the subsequent wave of information reflect quantitative precision as a central persuasive theme in science:

> The telescope [at Jodrell Bank] seemed to me to be ideally equipped to obtain much additional information about the pulsating source, particularly to extend the wavelength range . . . and the dispersion measurements. . . . Before the issue of *Nature* containing the announcement of the Cambridge discovery was delivered at Jodrell Bank on Saturday morning, the pulsating source was being received and studied on these higher frequencies. The amount of information which then flooded in, as soon as these advanced analytical techniques were brought to bear on the problem, raised great enthusiasm, and a week later two or three other individuals spontaneously abandoned their immediate tasks to help with the analysis and also to look for the signals with the MkII telescope in the llcm waveband. Whereas the Cambridge telescope could only observe the pulsating source for about 4 minutes per day, the Jodrell telescope could follow it continuously as long as it was

> above the horizon and consequently information about the pulse shape and fading characteristics accumulated rapidly.[22]

A less powerful, but useful, problem-solution *topos* is *empirical adequacy*. This *topos* is drawn upon when one argues that although a theory or body of data does not allow full comprehension of all the details of phenomena under study, the theory or data acquired to date reasonably account for the general nature of the phenomena. At least three material propositions are suggested by this *topos:* (1) The empirical adequacy of a theory will be demonstrated if there is adjustment of theory and observations so that the two are brought together in close agreement. (2) The adequacy of a theory will be justified if it is extended to cover areas to which it has not previously been applied. (3) The adequacy of a theory will become manifest with further collection of concrete data that justify the theory's application and/or extension.[23]

As an example, let us reconsider the arguments about the J phenomenon. Scientific rhetors were able to make refutative arguments against Barkla's claim by using the *topos* of corroboration. They contended that Barkla's experimental results were not confirmed by other experiments. As experimental evidence accumulated, the likelihood diminished that Barkla's theory would be substantiated. Hence, a reason for rejecting Barkla's research claim was that it was not empirically adequate. Its promise of future successful confirmation could no longer be accepted. On the other hand, if we agree with Wynne's assessment of Barkla's circumstances, we can see how conventional scientific theory was defended on the grounds of empirical adequacy. Wynne claims Barkla's radical claims presented received theory with important problems, but scientific rhetors and audiences decided it was reasonable to overlook those difficulties and to accept received theory as empirically adequate. Eventually, they thought, the problems could be worked out.[24] As this case shows, defenders of received theory can use arguments from the *topos* of empirical adequacy both to attack novel, challenging claims and to defend orthodox claims.

The *topos* of *significant anomaly* is resorted to when scientists argue about whether theoretical constructs are or are not in jeopardy in consequence of their capacity to explain, predict, or categorize new data. This line of argument occurs when findings violate expectations. The violation may be theory-threatening or it may be failure of a datum to be as small or large as expected. In mundane cases of anomalies the initial issue to be met will be, Is the anomaly significant enough to worry about at all? In more threatening circumstances the issue may be whether an entire theory or construct needs to be altered to accommodate anomalous findings. Arguments that urge or seek to repress intellectual revolution in a scientific community are, of course, extensively drawn from the significant anomaly *topos*. I have already shown how experimental parapsychologists and Francine Patterson tried to explain criticisms of their research as refusals to recognize and deal with significant anomalies allegedly possessing revolutionary significance.[25]

Scientific discoveries must often be justified with significant anomaly arguments. As Kuhn points out, at the inception of scientific discoveries one must not only become convinced *that* something has occurred but also that current categories and classifications cannot account meaningfully for *what* that something is.[26] The discoveries of oxygen, x-rays, and Uranus are cases in point, as is Bell's inability to classify the "scruff" on the recording charts. Bell testified, "I was interested in scintillation and I [knew] I had to reject man-made interference. This thing wouldn't quite fit into either category."[27]

Even when scientists admit anomalous data defy conventional interpretation, many lines of argument are available to counter claims that an anomaly is significant. It frequently is argued that the anomalous data are artifacts of experimental apparatus or design. Such arguments are drawn from the *topos* of experimental competence. One also can challenge anomalous data on grounds that they have not been confirmed by additional experimental or observational exploration (corroboration), or because experimental observations that identified the anomalies cannot be repeated systematically (experimental replication). Not least among available counter arguments is the contention that anomalies will be solved through additional experimental or theoretical work (empirical adequacy).

It is possible to establish reasonably that a significant anomaly confronts theory but does not overthrow that theory. As an example, consider the solar-neutrino anomaly. This anomaly emerges from a discrepancy between theoretical predictions and experimental data. Detection of neutrinos coming from the sun has been consistently and significantly lower than predicted by received theories of stellar energy generation. One astronomy textbook presented this anomaly as an "interesting problem," whose significance is amplified by posing alternative explanations that challenge received theory: the sun's interior is not undergoing nuclear reactions; neutrinos are not produced as predicted; knowledge about nuclear generation is wrong.[28]

Davis was the experimentalist at the forefront of identifying this anomaly in astrophysical theory. Pinch's investigation of how scientists in different specialties received Davis's experimental results is instructive about how the significance of anomalies becomes established.[29] As it turned out, factors pertaining to Davis's rhetorical ethos were most influential in inducing trust in his anomalous results. Pinch reports that many of his respondents referred to Davis's carefulness, modesty, and openness. These professional qualities contributed to forging consensus on the credibility of his results.[30] The consensus was formed largely through informal channels.

Closely related to Davis's professional virtues was constant and satisfactory examination of his experimental competence in producing credible results. Only one astrophysicist, Jacobs, challenged Davis's results in the open forum published in *Nature*. Jacobs contended that "the solar-neutrino problem is solely an artifact of the chemistry of the detection technique."[31] An article addressing Jacobs's criticisms appeared in *Nature*, and Davis himself satisfied most interested scientists by running experimental checks sufficient to ensure his results were not due to chemical trapping. Jacobs was never convinced, but his failure to gain tenure at the University of Virginia put an end to his criticisms.[32] In this case then, although it was not implausible for a scientist to challenge Davis's results, his results were judged reasonable because of his strong rhetorical ethos and because of other scientists' positive perceptions of his experimental competence. On these grounds the challenged theory and the anomaly were allowed to coexist.

The *topos* of *anomaly-solution* is used when one argues that a given claim solves a theoretical or experimental anomaly recognized as significant by a scientific community. For instance, during interparadigmatic conflicts the single most persuasive claim proponents of a new paradigm can make is that they can solve the problems that led the old paradigm into crisis. Kuhn has pointed out that much of the persuasiveness of claims by Copernicus, Newton, Lavoisier, and Einstein came from the fact that they appeared able to cope with recognized anomalies and so solve theoretical problems.[33]

Anomaly-solution arguments can also be made where there is no clash of paradigms. The method will still involve claiming that a certain way of thinking or observing or measuring should be replaced by another because that one better explains or clarifies a range of phenomena. This happens when scientists propose using different analyses or propose new methods of experimental control. The line and form of argument will be the same as when it is claimed that some new paradigm is necessary to cope with anomalous information.

EVALUATIVE *TOPOI*

Evaluative *topoi* suggest lines of argument in which rhetors test the special values of experimental, theoretical, or methodological claims. Lovell's account, cited earlier, decided the Jodrell telescope ought to give better data than the telescope used by scientists at Cambridge. The next natural question was evaluative: Better in just what ways? Lovell answered; "To extend the wavelength range . . . and the dispersion measurements." Evaluative claims that newly gathered data improved or extended the Cambridge data "flooded in," says Lovell.

Evaluative *topoi* suggest values that can be (and are) applied in choosing among theories or data or methods. Some values are inherent to empiricism, but some are more broadly based. In his various writings Kuhn identified several major values of scientific communities.[34] *Accuracy, internal consistency,* and *external consistency* are values empirically cogent work must have. Kuhn mentions additional values that do not derive necessarily from empiricism. *Scope, simplicity, elegance,* and *fruitfulness* are values that apply in a wide range of human activities. One could apply them as appropriately in evaluating an epic poem as in evaluating a scientific claim. From a rhetorical point of view all seven values are evaluative *topoi.* The first three are peculiarly productive of arguments made in science and formal logic. The last four yield arguments in science and in a wide variety of other realms. All, however, are evaluative *topoi* that suggest lines of argument and lines of logical critique. They recommend that a way of arguing about the credibility of claims is to consider their scientific value, whether intrinsically or in comparison with alternative claims. Kuhn's list of evaluative *topoi* can be expanded, but the list is sufficient to show the nature and use of evaluative *topoi* in making and judging scientific discourse.

Of all discussable matters scientists are least likely to accept or agree to theories, observations, instruments, or procedures containing inaccuracies. Satisfactory attainment of explanation, prediction, and control requires *accuracy.* Any scientific rhetor must therefore show that what he or she says ought to be taken as true methodologically and observationally: it is not casual, slipshod guesswork; it is the result of careful, appropriate analysis of data—it is accurate. However, a dispute between proponents of two theories is likely to hinge on the comparative accuracy of data and their interpretation. The comparative value of experimental claims may be defended or challenged in a variety of ways: by evaluation of laboratory preparations; by failure or success in applying standard procedures and methods; by precision or lack of it in identifying experimental outcomes; or by ambiguities in handling statistical calculations.

Accuracy is such a prime value in science that using that *topos* is almost automatic. Virtually all scientific claims are tacitly or explicitly undergirded by claims to accuracy. It is probably when an observation will surprise the community or when one is proposing a paradigmatic shift that strongest, most explicit emphasis on themes about accuracy will be required.

The *topoi* of *internal consistency* and *external consistency* are also compelling sources for scientific arguments. Both Ptolemaic (geocentric) and Copernican (heliocentric) astronomical theories were internally consistent. Thus, each could be justified as a good theory on the basis of arguments from this *topos*. However, the geocentric theory alone was consonant with the then current theories about related aspects of nature. As Kuhn observed, received physical theory needed the concept of a stationary earth to explain such phenomena as "how stones fall, how water pumps function, and why the clouds move slowly across the skies." Since heliocentric astronomy was externally inconsistent with respect to these other scientific explanations, this *topos* was of little persuasive value as a source of arguments on its behalf.[35]

The *topos* of external consistency becomes particularly important when scientific rhetors attack or defend aspects of a community's conventional knowledge. Orthodox scientists found the claims of experimental parapsychologists unreasonable because they contradicted currently accepted laws of physics. Similarly, Sebeok vigorously attacked Patterson's claims about language acquisition in gorillas because they were inconsistent with the received opinions of linguists. Terrace's claim that apes could not learn or interpret sentences was praised because it was consistent with established research conclusions in respectable scientific fields. In these instances external consistency was the consideration from which arguments were created and judged as "logically" reasonable.

It is not surprising that scientific rhetors striving to advance radical claims prefer to defend their research with arguments drawn from the *topos* of significant anomaly. They explain away orthodox opponents as spirited defenders of a decaying intellectual order in crisis. For example, when they were unable to meet the charge that their claims were inconsistent with received knowledge, Patterson and several parapsychologists argued that they were bringing to light fundamental problems in the community's conventional knowledge. They met charges of external inconsistency by contending that orthodox knowledge was beset by significant anomalies and hence was itself inconsistent with empirical facts.

In each of the above instances those under attack tried to change the point of *stasis* in the controversy. Instead of external consistency they wanted to put significant anomaly at issue. However, parapsychologists and Patterson continue to be persuasively disadvantaged because in the scientific communities they have to address, there remain presumptions in favor of existing formulations. Conceivably events could alter the accepted paradigms and admit the parapsychologists' and Patterson's claims. We know this happened in the controversy between Ptolemaic and Copernican ideas. In any case persuasion, not brute fact, wins the day in a good many such disputes.

Arguments about external consistency draw much of their persuasive power from what Quine and Ullian called conservatism, the reluctance of scientific communities to disturb existing beliefs.[36] This is not insidious. It is a specific appearance in scientific activity of a general rhetorical principle called presumption. In his ninteenth-century ecclesiastical treatise, *Elements of Rhetoric,* Richard Whately defined this principle by comparing it with the corollary idea of burden of proof. Presumption does not mean "a preponderance of probability" in a claim's favor, "but, such a *pre-occupation* of the ground, as implies that it must stand good till some sufficient reason is adduced against it; in short, that the *Burden of proof* lies on the side of him who would dispute it."[37] Conservatism in science simply reveals the presumption which Whately identified as standing against anything contrary to prevailing opinions.[38] Conservatism is a tendency that makes arguments from the *topos* of external consistency particularly persuasive.

A good scientific theory should have broad *scope*. Theories are tested and defended as to whether their consequences extend beyond particular observations, laws, or subtheories. In other words, the value against which a theory is weighed is comprehensiveness. Consider the initial response to early memory-transfer studies with rats. As I have pointed out, some worm studies had produced the conclusion that chemical memory transfer had occurred. This violated the orthodox theory that memory involved networks of nerve cells, but the anomaly did not receive much attention until it was claimed that memory transfer occurred in rats. The implications of the claim now reached to animals whose memory operations were thought to be similar to humans'. As Travis put it: "As a phenomenon associated with worms, memory transfer *could* be regarded as a biological oddity of limited relevance and applicability. Claiming that the phenomena operated in rats was far more consequential. So, despite a knowledge of the worm-running experiments, the first mammalian results produced strong reactions."[39] They had broader scope.

The *topos* of *simplicity* also may be used to justify a theory or claim. It is a commonplace of science that parsimonious explanations are best. The central line of thought is that what can be done with few assumptions is done in vain if more are injected. Accordingly, rhetors' claims are strong if they involve a minimum of assumptions and conditions, and in some cases even a minimum of empirical data, provided data are clear, uniform, accurate, etc.

Debates about the alleged capacity of apes to acquire language underscore the general situational availability of parsimony as a *topos* for scientific argument. In a critical review of the literature Umiker-Sebeok and Sebeok observe that "the most common form of criticism leveled by one ape 'language' investigator against another is that simpler, more parsimonious explanations are available to account for a rival's data."[40] As an example, they show how Savage-Rumbaugh et al. apply this *topos* refutatively:

> Overly rich interpretations of chimpanzee and gorilla ASL [American Sign Language] productions abound in the published reports. Savage-Rumbaugh *et al.* have repeatedly argued that, given the weak acquisition criteria employed by those working with signing apes—namely, the ability to name pictures or to produce an iconic gesture repeatedly, under the same or similar circumstances—"it is impossible to tell whether the chimpanzee is simply imitating or echoing, in a performative sense, the action or object, or whether the animal is indeed attempting to relay a message." Since, as they further allege, ASL is a highly iconic system of communication, they conclude that there is nothing to prevent us saying parsimoniously that the apes are merely producing "short-circuited iconic sequences." Unless it can be shown, they go on to say, that the ape could competently use a sign apart from the original context in which it was acquired (i.e., in a situation where the actions and signs of the trainer cannot remind the animal of the hand movements used in the sign), the mere announcement that signs occurred in "new contexts" is not sufficient to rule out the possibility that the animal's behavior

> can be accounted for by the simpler explanations of deferred imitation or error interpreted in a novel manner by the experimenters.[41]

Parsimony is logically used in either refuting or establishing claims. Umiker-Sebeok and Sebeok use the *topos* to support their case against interpretive claims attributing intentionality to apes' signing. For instance, they argue that allegedly innovative signing behaviors are more parsimoniously explained as mistakes or responses to experimenters' cuing than as jokes, insults, or metaphors.[42] Indeed, they use parsimony to insinuate that Savage-Rumbaugh et al. are themselves as guilty of using "richly interpreted anecdotes" in their work as the language investigators they seek to criticize.[43] In contrast, Griffin used this same *topos* to support the case *for* interpretive claims that Savage-Rumbaugh et al. put forward regarding signing behaviors of two chimpanzees, Austin and Sherman. He argued it was more parsimonious for Savage-Rumbaugh et al. to explain those behaviors as intentionally communicative than to adopt alternative explanations put forward by critics of ape language research generally, and Sebeok specifically:

> The ability of Austin and Sherman to communicate about tools needed to acquire particular foods provides important new evidence that chimpanzees can communicate intentionally with at least rudimentary understanding of what they are doing. There will doubtless be some, like Sebeok (1977), who continue to suspect that unrecognized cues, *un*intentionally provided by experimenters, may be responsible for what looks so much like intentional communication. But the complexity that would have to be postulated in such "Clever Hans" explanation makes the assumption of intentionality seem parsimonious.[44]

The issue about whether parsimonious explanations should include or exclude the assumption of intentionality in respect to apes' behaviors underscores an important point about the logic of arguments derived from all evaluative *topoi:* No true-false conclusion is or ever can be established by arguments from the *topos* of simplicity or any other evaluative *topos*. The logic of scientific evaluation only allows judgments about the legitimacy of claims *relative* to other claims and the situationally derived criteria that purportedly legitimate claims. Once more we see that a logic other than formal logic legitimates scientific discourse.

An evaluative *topos* closely related to simplicity is the *topos* of *elegance*.

This *topos* gains its persuasive force from scientists' aesthetic sensibilities and judgments. One theory may be judged neater, prettier, or better balanced than another. Logical or mathematical formulations are considered elegant in contrast to cumbersome alternatives. A given set of facts may be said to fit neatly with what is already known. That is a persuasive virtue. Bertrand Russell asserted the persuasive power of arguments from this *topos* in his eloquent discussion of aesthetic appeal in mathematics:

> Mathematics rightly viewed, possesses not only truth, but supreme beauty—a beauty cold and austere, like that of sculpture. Without appeal to any part of our weaker nature, without the gorgeous trappings of painting or music, yet sublimely pure and capable of stern perfection such as only the greatest art can show. The true spirit of delight, the exultation, the sense of being more than man, which is the touchstone of the highest excellence, is to be found in mathematics as surely as in poetry. What is best in mathematics deserves not merely to be learnt as a task, but to be assimilated as part of daily thought and brought again and again before the mind with ever-renewed encouragement.[45]

Scientists consider aesthetic appeal when appraising substantive theories as well as strictly formal claims about logic and mathematics. The decline of memory-transfer research with mammals is a case in point. Rose observed that molecular-memory models were rejected as scientifically unreasonable partly because they lacked aesthetic value: "At their crudest, the models of memory mechanisms which depend on molecules rather than structures not only run counter to the conventional wisdom of neurobiology, but *they are also unaesthetic, a matter of some significance for the acceptibility of scientific theories.*"[46] Whether used to support or reject claims, arguments drawn from this *topos* occur repeatedly in scientific discourse and are avowedly intended to be persuasive.

Fruitfulness is a *topos* that suggests premises from which to argue that a claim will be productive of new findings, will expand existing theory, or will otherwise make for progress. If a claim is shown to disclose new phenomena or new relationships among phenomena, that can be offered as a persuasive strength. Furthermore, a theory that is established as fruitful may be especially persuasive to research scientists because of implications for their research careers.[47] A particular datum or array of data can be made more interesting, credible, and appealing if it can be shown that incorporating the data into current understandings renders a pattern strong enough to justify further research or theorizing.

When Hewish and the Cambridge group announced the discovery of pulsars in 1968, their claim was immediately seized upon as potentially fruitful. (A part of what happened is told in the quotation from Lovell cited earlier.) However, the Cambridge group mentioned three other pulses in their discovery announcement, but they did not release the coordinates. According to Lovell at Jodrell Bank, Hewish and his colleagues wanted to keep this information secret. An independent search for the three sources would have been time-consuming. Other research groups put pressure on Cambridge to release the coordinates because investigation of these three other sources and comparisons of them with the one already investigated would, in a word, be fruitful. When the coordinates were finally released, Jodrell Bank scientists focused their efforts on investigating the phenomena and, in Lovell's words, "there was an outpouring of results such as occurs only rarely in the life of a scientific establishment."[48] There is no doubt that scientists found the data on pulsars valuable because they were fruitful for solving problems the community thought were of special importance at this time.

New theories, new interpretations, and new data are seldom as self-evidently fruitful as the pulsar findings. Then scientific rhetors have to invent arguments showing that what they offer is actually or potentially fruitful. The maker of such arguments need not be the person who offered the findings. Scientist A may report data and scientist B may be the one who argues that, given this or that interpretation, the data are more fruitful than A believed or noticed. In all these cases fruitfulness, as a *topos,* identifies lines of argument that can be followed in the rhetoric of science.

EXEMPLARY *TOPOI*

The third class of rhetorical *topoi* that scientists use in arguing about allegedly significant knowledge claims are exemplary *topoi.* Exemplary *topoi* include such discursive strategies as examples, analogies, and metaphors. Exemplary *topoi* are most nearly like the general rhetorical topics discussed in chapter 5. The concepts—example, analogy, and metaphor—suggest lines of thought for making persuasive discourse in any field and in any rhetorical situation. Such resources are never out of place. Nevertheless, when these *topoi* are specially adapted for use in invention of scientific rhetoric, they have special rhetorical functions.

Perelman and Olbrechts-Tyteca made the important point that rhetorically effective uses of examples, illustrations, analogies, and metaphors assert tacit or explicit claims about the structure of reality.[49] All of what I call exemplary *topoi* posit certain relations in the structure of reality. They all suggest that some ordering rule transcends the particular example, illustration, analogy, or metaphor used. To use a hypothetical example, if I want to clarify the difference between a compound and an element, I can instance brass—a compound of copper and zinc (illustration); or I can compare compounding to scrambling eggs and milk to make scrambled eggs (analogy); or I can speak of a compound as a marriage of elements (metaphor). In each case I draw from the exemplary *topos* a way of thinking that posits the general rule that all compounds are intermixtures of elements.

Exemplary *topoi* identify rhetorical possibilities in science as they do in other rhetoric, but in science the exemplary *topoi* function differently in investigation and in rhetoric about findings. Commentators on scientific activity have struggled with the problem of defining the function of what I am calling exemplary *topoi* in the investigative and imaginative dimensions of scientific inquiry.[50] Generally, exemplars function to conceptualize phenomena in ways that guide investigations. For instance, Kuhn's notion of exemplars can be thought of as a heuristically powerful kind of example. They are a scientific community's concrete achievements, and thus demonstrate how significant problems have been solved. Paradigmatic exemplars suggest fertile lines and techniques of investigation that can be used to bring similar problems and related puzzles to solution. Black's discussion of theoretical models resonates with lines of thought characteristic of the exemplary *topoi* of analogy and metaphor: Heuristic models supply useful analogical fictions that help scientists probe and conceptualize relationships among phenomena; ontological or existential models prescribe metaphysical commitments about what phenomena really are. Black clarifies the difference between heuristic and ontological models with the example of Maxwell's representation of an electrical field in terms of the properties of an imaginary incompressible fluid.

When Maxwell first accounted for his modeling procedure he emphasized that it was no more than a heuristic device for grasping mathematical relations that could be expressed more precisely by algebraic equations. The fluid was, after all, "imaginary." In later discussions Maxwell revealed a growing ontological commitment to his image. He spoke of a "wonderful medium" filling all space, and that Faraday's lines of force "must not be regarded as mere mathematical abstractions. They are the directions in which the medium is exerting tension like that of a rope, or rather, like that of our own muscles." Thus, as Black points out, "the purely geometrical medium has become very substantial."[51] In a succinct statement of the distinction between a heuristic and an ontological model Black maintains that

> the difference is between thinking of the electrical field *as if* it were filled with a material medium, and thinking of it *as being* such a medium. One approach uses a detached comparison reminiscent of simile and argument from analogy; the other requires an identification typical of metaphor.[52]

Exemplary *topoi* are used both in investigation and in making rhetoric about investigations. In either case one suggests to oneself that examples, analogies, and metaphors might be useful. However, when one goes to these inventional resources in order to make rhetoric about science, one needs to choose examples, analogies, and metaphors for their communicative and situationally clarifying virtues. This often requires that one choose exemplary *topoi* that reflect a community's special research interests. Pickering's point about special groups having an interest in furthering research based on a shared exemplary paradigm has rhetorical implications:

> An exemplar is an example *for some particular group*—the group which has established the preceding body of practice—and not for others. One can speak of the group or groups having expertise relevant to the articulation of some exemplar as having an "investment" in that expertise, and as a corollary, as having an "interest" in the deployment of their expertise in the articulation of the exemplar.[53]

Choosing rhetorically useful exemplars must be done in view of community research interests which constrain the range of technically acceptable exemplars for that community. Reasonable claim-making cannot be accomplished unless selected examples, analogies, and metaphors intersect with situationally relevant research orientations (and corollary interests) that targeted audiences are using to comprehend the socio-logic of technically "legitimate" problem-solving.

When a scientist assumes the stance of rhetor, the ideas and arguments associated with exemplary *topoi* also open up communicative possibilities far beyond those allowed by the strictly technical considerations associated with communally prescribed uses of scientific method per se. Examples, analogies, and metaphors can also be chosen to arouse a particular audience's interest. Put differently, as a scientist's task shifts from simply investigating to making the results of investigations public, social constraints in addition to technical expectations can enter into choices of what to say. Some questions among those pertinent in making such choices are: What image will most precisely but also most colorfully express the scientifically justified notion I have? Of all examples available, which will most quickly grasp the attention and interest of this audience? What analogy is both precise enough to be scientifically acceptable and interesting enough to seize and hold attention? These and related questions become especially important when rhetoric about science is composed for popular consumption. Lewis Thomas is a master at making rhetorical choices without sacrificing scientific integrity when composing discourse about science for general audiences. For example, in *Lives of a Cell,* when Thomas wanted to express and support his claim that in lower organisms communicative behavior is specific, ideally adapted to function, but inflexible, he wrote:

> "At home, 4 p.m. today," says the female moth, and released a brief explosion of bombykol, a single molecule of which will tremble the hairs of any male within miles and send him driving upwind in a confusion of ardor. But it is doubtful if he has an awareness of being caught in an aerosol of chemical attractant.[54]

Thomas discovered (1) an *example* of communication among lower organisms (the moth), (2) an appropriate but not absolutely necessary *metaphor* with which to express communicative behavior (courtship), and (3) an *analogical contrast* between degrees of awareness in lower organisms and in humans. Two important points must be made about this scientist's use of exemplary *topoi.* First, simply to assert his basic claim Thomas needed no example, analogy, or metaphor. Second, he could have chosen any number of ways of exemplifying and portraying the operation of his claim. His exemplars are rhetorically excellent because they are appealing and mildly amusing; they are interesting because set in tiny narrative; and they call up sexual images and associations—always intriguing to humans as humans. Used rhetorically examples, analogies, and metaphors lead to formulations that interest and simplify while being at least suggestively accurate.

Thomas's rhetoric about science also illustrates why it is useful to pull together under the single heading of exemplary *topoi* the otherwise disparate lines of argument designated by examples, analogies, and metaphors. Any exemplary *topos* used rhetorically always implies the existence of some rule or law that applies beyond the instances cited. The rule may be stated or unstated. It is not stated in the passage quoted from Thomas, but one can grasp the principle of deterministic, unchoosing communication from the passage. In his book, of course, Thomas stated his thesis directly; but the general point is, if exemplary rhetorical *topoi* yield clear and persuasive examples, analogies, or metaphors, those devices will always stand as exemplars of how-things-are, either generally or under specifiable circumstances. This is true regardless of whether the rhetor is seeking to address a scientific or nonscientific audience.

Each kind of exemplary *topos* yields a different way to think about how-things-are despite the fact that all such *topoi* implicitly or explicitly assert a transcendent rule or principle. Perelman and Olbrechts-Tyteca assert that rhetors use examples when a general rule or principle is somehow at issue.[55] *Examples* are useful for establishing a rule or principle. In contrast, *illustrations* (a subtype of example) provide additional support for, or clarify, already established regularities.[56] For instance, brass served as a particular illustration of compounds. This illustration clarifies and, to some extent, increases adherence to an already established principle of ordering even though it does not help to establish the reality of that structural principle. Perelman and Olbrechts-Tyteca explain that in scientific discourse "particular cases are treated either as examples that are to lead to the formulation of a law or the definition of a structure or else as specimens or illustrations of a recognized law or structure."[57]

Analogy can establish or clarify one structure by positing that it resembles another more familiar one. It follows the form: A is to B as C is to D.[58] Analogies work rhetorically when audiences judge that structural principles governing relationships in one sphere are similar to those governing the second sphere. Thus the principle is asserted that compounding is achieved by combining elements in a way that structurally resembles making scrambled eggs by mixing eggs and milk. Certainly analogies have important expository uses in scientific education and technical training, but they can also supply strategies of argumentation that stress unforeseen similarities between relationships found in an orthodox scientific specialty and structures found in another specialty, or even in the general culture. Using such a strategy, intellectual revolutionaries seek to reorient the thinking of audiences, to shift and transform ways of thinking about and discussing problems in a specialty.

Analogical comparison is a common line of argument in scientific discourse. Much of what is said in the experimental literature in behavioral psychology depends for its rhetorical effectiveness on some degree of adherence to the idea that rat and human physiology and behavior are sufficiently similar to warrant inferences from experimental results with laboratory rats to conclusions about humans.[59] Similarly tests of chemical toxicity using laboratory animals are thought to provide guidance about determining product safety for humans. This kind of argument can be persuasive only if it is tacitly accepted that under specifiable conditions animal and human physiology are sufficiently similar to warrant drawing generalized conclusions applicable to both species.

The exemplary *topos* of *metaphor* yields another strategy for establishing or clarifying a structural principle. It brings together connotations and associations drawn from different spheres of thought. Along with analogy this *topos* can encourage lines of thought found deep in the wells of creativity, scientific or other. The power of metaphor is not captured by considering it merely ornamental, as some literary critics and some traditional rhetorical theorists have suggested.[60] Metaphors are not mere pretty packages of content that could be expressed in plainer and clearer terms; they are, as I. A. Richards contended, the very instruments of thought.[61] Black has explained the creative processes involved:

> A memorable metaphor has the power to bring two separate domains into cognitive and emotional relation by using language directly appropriate to the one as a lens for seeing the other; the implications, suggestions, and supporting values entwined with the literal use of the metaphorical expression enable us to see a new subject matter in a new way. The extended meanings that result, the relations between initially disparate realms created, can neither be antecedently predicted nor subsequently paraphrased in prose. We can comment *upon* the metaphor, but the metaphor itself neither needs nor invites explanation and paraphrase. Metaphorical thought is a distinctive mode of achieving insight, not to be construed as an ornamental substitute for plain thought.[62]

Through metaphor we identify an ordering principle by transferring the meanings of customary usage to another logically different usage or conception. The interaction of the two complexes of meaning brings together similar ideas and images within each, thereby creating, establishing, or clarifying a structural principle. However, when metaphor is put to use as a rhetorical strategy, additional concerns must be addressed.

"Marriage of elements" is a metaphor for compounding. Some connotations and associations that the concept "marriage" calls up will serve a rhetorical purpose of clarifying and vivifying, but some others will not.

If the metaphor is to work as I wish, my respondents must transfer such notions as intimate union, consummation, uniting together, and other conceptions of joining individuals together in a fixed and formal way. My metaphor will fail rhetorically should other associations of marriage arise in respondents' minds: honeymoons, divorce, creating a family, and the like. These are inimical to the image of bonding that I wish to evoke. From a rhetorical point of view my metaphor functions as the right kind of argument and clarification if, culturally and situationally, my particular audience is more likely to associate marriage with bonding than with other inimical dimensions of the metaphoric concept. In scientific rhetoric, as in all rhetoric, metaphors need to be chosen for their strategic potentialities, given the audience addressed.

Metaphors possessing special rhetorical power engage commonplace understandings and associations while narrowing or broadening perceptions in technically meaningful ways. This sometimes allows for their use in both popular commentaries and in technical discourse. The "communication" system of metaphor used to discuss the DNA molecule's genetic functioning is a case in point. Halloran and Bradford observe that popular discourse is replete with references to "a genetic 'code' in which the DNA molecule 'transmits' 'information' that is 'transcribed' onto the substances in the cell and thus 'translated' into the characteristics of a specific organism."[63] In the technical literature of molecular biology they find that "terms, such as 'expression,' 'translation,' 'transcription,' 'messenger,' even 'editing' and 'reading' are used to describe the working of DNA."[64] The "message" system of metaphor is rhetorically rich because it has strategic potentialities with different kinds of situated audiences. For scientists it invokes familiar associations for focusing perceptions in technically legitimate and heuristically fertile ways; for the laity it exposes common grounds for assimilating foreign technical ideas into familiar cultural notions.

Scientists can of course disagree about whether associations that are metaphorically induced broaden or narrow perceptions in scientifically reasonable ways. When they do, arguments about the choice of metaphors will frequently result. The charge is sometimes made that those using a metaphor have forgotten that it is based on a suggestive series of analogical relationships and should not be confused with literal expressions of "real" structures in the world.[65] This charge is usually implicit in accusations of reductionism of one kind or another. For instance, those using machine metaphors to depict human brain functioning (e.g., the brain is an electronic digital computer) have been accused directly of mechanistic reductionism.[66] Insinuation of anthropomorphic reductionism can be made by drawing attention to the metaphorical nature of terms about humans when they are used to make literal interpretive claims about animal behavior. Umiker-Sebeok and Sebeok's parenthetical remarks thus subtly work toward diminishing the reasonableness of Patterson's interpretive claims about the gorilla Koko's allegedly innovative sign behavior: "It was interpreted as a deliberate joke (the gorilla was 'grinning') when Koko, in response to persistent attempts to get her to sign 'drink,' made 'a perfect drink sign—in her ear.' "[67] When treated literally, a "human" system of conceptions about animal behavior (joking, grinning) can expand or restrict associations in scientifically unreasonable ways.

Scientists make frequent use of exemplary *topoi* when trying to establish an ordering principle that presumably structures how-things-are in a particular field of phenomena. As a case in point I shall consider portions of the debate about cerebral localization conducted during the late nineteenth century, with special attention to the arguments of Sir David Ferrier and Friedrich Goltz. This debate illustrates how the exemplary *topoi* of example, analogy, and metaphor were used to establish and to challenge the principle that motor and sensory functions can be localized in distinct regions of the brain.[68]

The central point at issue was whether the brain's structure could be differentiated according to regions that were "centers" for observable behaviors. Localizationists would choose evidence that allowed them to argue for the existence of such regions as a "speech area" or a "leg area." Antilocalizationists embraced the idea of cortical equipotentiality which, rather than differentiating the brain into special functional regions, asserted that the brain functions as a whole, with any part of the cortex potentially able to take over and discharge the functions of other parts. Accordingly they chose evidence allowing them to posit explanations based on more integrative reactions of the brain as a whole. Both sides supported their claims with arguments by example.

Localizationists conducted experiments that sought to match overt behavioral functions with specific regions of the brain. A common practice was to stimulate electrically a circumscribed area of an animal's exposed cortex, observe behavioral responses that were thereby induced, and interpret the results as indicating the existence of, say, a "motor area" for the right forelimb. To clinch their case experimenters would then surgically remove the electrically defined forelimb "center" and interpret subsequent loss of ability in using the forelimb as evidence for a forelimb center.[69] Results from several such experiments, each focusing on different regions, could then be cited as examples establishing the general principle of localization.

Sir David Ferrier made persuasive use of arguments by example in his historic debate with antilocalizationist Friedrich Goltz at the 1881 International Medical Congress in London. Ferrier presented a series of cases drawn from his experiments with monkeys.[70] Each demonstrated how injury to a specific brain area resulted in loss or severe impairment of the animal's ability to execute a specific behavioral function. Damage to one area resulted in blindness, thus demonstrating a vision center. Damage to another area made the animal deaf, thus establishing a hearing center. Injury to other portions of the brain impaired an animal's ability to move its right arm and right leg, thus showing the existence of special centers for those specific motor functions. Ferrier concluded that these and other examples "are sufficient to demonstrate that Professor Goltz's hypothesis is erroneous, and that such facts are explicable only in the theory of distinct localization of faculties in definite cortical regions."[71]

Goltz also used experimental cases to make his own arguments by example. He removed parts of dogs' brains that were thought to be motor and sensory centers. Though these animals suffered a "general lowering of intellect," they eventually recovered ability to perform those specific motor and sensory functions. Goltz cited such cases as examples for establishing cortical equipotentiality and refuting localization. Any behavioral disturbances were interpreted as the reaction of the brain as a whole to the *magnitude* of damage inflicted, rather than the precise location of lesions.[72]

The adjudicating audience decided in favor of Ferrier's examples exemplifying localization. Central to deciding the issue were the comparative strengths of the two men's demonstrations using live animals. Goltz put a dog through various tasks demonstrating that although the animal was, in his words, a "canine imbecile," it still possessed motor and sensory abilities that should have been extinguished after Goltz allegedly had removed the relevant "centers" in the brain. Ferrier demonstrated how he had surgically caused one monkey to be deaf and another to lose motor ability in its right arm and right leg, attributing these functional disturbances to extirpated "centers" in the brain. Both the dog and the motor-impaired monkey were then sacrificed and their brains were extracted for examination. A committee of examiners found that placement of injuries to the monkey's brain corresponded exactly with Ferrier's claims, but that Goltz failed to remove all sensory and motor areas as he had claimed.[73]

Analogical comparison was also used by controversialists in the cerebral localization debates. For instance, once examples were provided to support the existence of a "leg center" for monkeys, it was argued by analogy that a similar spot in the human brain designates the region for the same motor function. Ferrier used this kind of reasoning when constructing his highly influential cortical maps of the human brain. Given the virtual absence of experimental evidence about human brain functioning, Ferrier used his maps of motor and sensory centers in monkeys' brains analogously when plotting the specific places for those functions on maps of the human brain.[74] For such an argument to work, an audience must grant that the functional relationship between a monkey's brain and, say, the motor behavior of its leg is essentially similar to that found in a human brain and the operation of a human leg.[75] Goltz had been criticized for extrapolating his results with dogs to human brain physiology, but it seems adjudicating audiences were more willing to grant the situational reasonableness of Ferrier's analogical comparisons, though those comparisons would ultimately be rejected as empirically inaccurate.

Ferrier prefaced his analogical comparison between functional centers in monkeys' brains and like centers in human brains with acknowledgement that "exact correspondence can scarcely be supposed to exist."[76] He nonetheless had suggested that similarities were sufficient to warrant plotting "regions" on human cortical maps, making direct analogies from experimentally determined places on cortical maps of monkeys. He sought to maximize the scientific reasonableness of this comparison by mentioning the gross structural similarities between monkey and human brains, and by asserting that evidence from human cerebral pathology and from a single experiment involving a human subject empirically supported some plotted centers.[77]

Ferrier's comparison was also bolstered tacitly by situational constraints that were favorable to the claims he was making. He could implicitly rely on evolutionary notions as justifications when he published his influential book in 1876. Indeed, Ferrier used the idea of evolutionary hierarchy to construct convincing analogies in his debate with Goltz. "Higher" animals (like monkeys and humans) share a comparable likeness in possessing differentiated brain functions, while "lower" animals (like frogs, pigeons, and Goltz's dogs) share in common lesser differentiation of brain functions.[78] This point contributed to Ferrier's persuasive success in the debate.[79] His analogies and demonstrating examples seemed, comparatively, more situationally reasonable than Goltz's.

An important social factor that may also have contributed to the situational reasonableness of Ferrier's analogical comparisons is that localization theory seemed to possess much greater promise for clinical application than opposed theories. He and other clinicians were eager to use the theory in diagnosing and surgically treating brain diseases.[80] Ferrier's analogical comparisons between monkeys and humans turned out to be empirically inaccurate, but the explicit backing he offered together with implicit situational constraints contributed to judgments that it was reasonable for Ferrier to assume the relationship between monkeys' brain centers and corresponding motor and sensory abilities was sufficiently similar to humans to warrant the comparisons.

The concept of cortical maps also had subtly persuasive influences in the localization debate. Maps of functional cortical areas were published. One rhetorical consequence was that by portraying the results of experiments as indicating positions on a map, it was argued implicitly that investigators were "filling in" preexisting, functionally differentiated "regions" of the brain.[81] Questions about whether specific functional regions or centers in fact existed could often be begged. This was a rhetorical disadvantage for antilocalization scientists, however right their thinking may have been.

"Mapping" did not cause excessive efforts at localization, but it is worth observing that a map system of metaphor generates associations that reinforce what critics of localization thought was its all-too-narrow view of the brain's operation. Maps invoke notions of boundaries or regions, and thus center attention on differentiation rather than integration. Antilocalizationists were concerned that integrative, interactive, interconnective processes involved in the brain's operations were obscured by localization. Maps represent surface territory but conceal subterranean depths. Even some mappers granted cortical maps had this deficiency.[82] The map metaphor alone should not be blamed for the excesses and limitations of cortical localizationists, but we should notice that conceptions associated with mapping could and apparently did promote and reinforce narrow and often faulty thinking about human brain operations. In science, as elsewhere, metaphoric thought can confine as well as free the minds of rhetors and their audiences.

CONCLUSION

The three kinds of *topoi* I have identified suggest durable ways in which scientific communities think and value. As such they index accepted, logical structures in scientific discourse. Resort to any of them may lend legitimacy to claims. They are structures of acceptable reasoning used over and over in scientific discussions. In the abstract they are at least ways of thinking that need not be defended as bases for arguments because all scientific communities accept and value them and, indeed, directly or indirectly teach them as features of scientific methodology.

There can, of course, be disagreement about the appropriateness of applications of any of these *topoi.* The accuracy of a metaphor, the closeness of an analogy, or the representativeness of an example can be challenged. The comparative importance and relevance of any *topos* will also vary with particular rhetorical situations.

These rhetorical *topoi* identify ways of making situationally logical, reasonable claims and comments. For creating or evaluating scientific discussion these classes of thought structures exist by virtue of standards that all scientists apply in doing science. I have not explored *topoi* that are peculiar to some scientific communities and not to others, but these could be isolated in the same manner that I have used in extracting *topoi* used in all sciences. We know that engineers and social scientists concern themselves with social utility whereas mathematicians do not, and that architectural engineers concern themselves with aesthetic/structural balance in ways that chemical engineers do not. Close analysis of discourse within scientific specialties would show us what other *topoi* are peculiar to each science.

The general points I wish to make are (1) that there are in fact identifiable lines of thought that are used again and again in the sciences; (2) that these lines of thought legitimize scientific observations and claims because they derive from what is accepted and valued in scientific communities; and (3) that if we want to see what the logical formulas and characteristics of scientific discourse are, we must grant that these *topoi* identify structures of thought that scientists (and often others) find situationally reasonable. These themes constitute a stable, ever-present collection of discussable options that are, as it were, culturally "authorized" by all scientific communities. Knowing what these options are is helpful to any scientist deciding what to say and how to say it to his or her colleagues. Knowing them is also useful in critique of scientific rhetoric because the array of standard *topoi* will remind observers of what *could* have been said about a scientific subject. With knowledge of these options a creator or critic of scientific rhetoric is able to estimate the relative usefulness or nonusefulness of themes in specific rhetorical situations. Knowledge of legitimated themes is also knowledge of what *topoi* will not count as reasonable, scientific thought; for example, the *topos* of "inspiration" is not in the array of *topoi* authorized by scientific communities.

The types of *topoi* discussed in this chapter probably do not exhaust the kinds of recurring logical structures of scientific discussion. They are, however, the kinds of themes that have recurred most often since the so-called scientific revolution. More detailed examination of the rhetorical practices of scientists may uncover additional, universal classes of scientific argument, and such exploration can undoubtedly enable scholars to identify the standard lines of thought peculiar to any specific scientific field.

10

PRACTICING RHETORICAL INVENTION: Creating Scientifically Reasonable Claims

Rhetorical inducement of scientific audiences succeeds or fails depending on whether scientific rhetors render their claims reasonable, according to the judgments of their situationally constrained audiences.

My claim is that analysis of the rhetorical features of scientific communication can reveal aspects of doing science that are not otherwise readily perceived. The kind of rhetorical analysis I have proposed in this book has primarily to do with the making of scientific communication. I do not suggest that this is the only significant aspect of carrying out scientific projects. I do believe, with McMullin, that beyond the formally logical "face" of science there is an "interpretive face," but I have further contended that whenever scientific communication is made, it brings into prominence situationally logical features of what can best be distinguished as the *rhetorical* face of science. Some have been unable to recognize this rhetorical dimension in science, seeing rhetoric instead as something irrational or illogical because it is fundamentally groundless, constrained by only the most idiosyncratic and subjective standards.[1] Other rhetorical dimensions of scientific communication such as arrangement, style, and presentational form have been explored in at least preliminary ways by others.[2] I have shown that close attention to *inventional* decisions reveals the informally logical dimensions of science that govern scientific reasonableness. The controlling logic of scientific discursive practices turns out to be topical rather than formal.

My objective in this chapter will be to illustrate further the kinds of observations and evaluations one can derive from examining inventional choices displayed by scientists and others responding to rhetorical situations in which the themes and values of doing science apply. I shall analyze the rhetorical choices made in two quite different kinds of rhetorical situations.

The first case I shall consider is the rhetoric made in and in relation to *McLean v. Arkansas*. In this case the reasonableness of giving balanced treatment to creationism in science curricula was at issue. The rhetoric associated with this trial illustrates the inventional choices scientists and purporting scientists are likely to make in rhetorical situations where the prime exigence is to demarcate what is and what is not science and where the decision is to be made by nonscientists authorized to adjudicate what does and does not count as science. The second case I shall analyze concerns Watson and Crick's article announcing the bihelical structure of the DNA molecule. The rhetoric associated with this article is prototypical of the inventional decisions scientists make when trying to render theoretical claims reasonable and persuasive for other authoritative scientists. I hope to show (1) the practical usefulness of rhetorical theory in criticism of scientific discourse, and (2) that degrees of scientific reasonableness will vary according to (a) the availability to a rhetor of appropriate (i.e., scientific) purposes, issues, and topics, and (b) the rhetorical wisdom of rhetors' choices from among available ideas. The two cases have special utility for these purposes because they illustrate how scientists practice rhetorical invention when seeking to establish reasonableness in what we may call "lay" and "technical" rhetorical situations.

CREATIONISM AND THE RHETORIC OF SCIENCE

The *McLean v. Arkansas* case came about in the following way. With the signature of Arkansas Governor Frank White, Act 590, the Balanced Treatment for Creation–Science and Evolution–Science Act, became law on March 19, 1981. The law called for equal time in science classrooms for presenting creationism and evolutionism as models of how the universe came into being and how life came to be. On May 27, 1981, a group of twenty-three Arkansas citizens and organizations brought suit in U.S. District Court charging that Act 590 was a disguised attempt to establish religion in the public schools and therefore was in violation of the First Amendment of the Constitution. The trial of the Rev. Bill McLean et al. versus the Arkansas Board of Education began on December 7, 1981, and lasted two weeks. On January 5, 1982, Judge William R. Overton expressed the Court's opinion that the act was unconstitutional on the ground that it had violated the Establishment Clause of the First Amendment of the Constitution.[3]

In seeking to render a reasonable decision, Judge Overton had to secure both legal and extralegal standards. He needed meaningful legal criteria to apply the Establishment Clause to the issues. For these criteria, he turned to precedent, a master *topos* of legal rhetoric, to find lines of thought with which to form a decision that would seem legally reasonable. At the very outset of his opinion he cited several precedents which, in his expert view, clarified the meaning of the Establishment Clause. Among the precedents cited was *Stone v. Graham,* a case adjudicated by relying on a three-pronged test for constitutionality in establishment cases. The test, used earlier in *Lemon v. Kurtzman,* was: "First, the statute must have a secular legislative purpose; second, its principal or primary effect must be one that neither advances nor inhibits religion; finally, the statute must not foster 'an excessive government entanglement with religion'."[4] Judge Overton turned to such precedents to establish noncontingent, noncontroversial standards for resolving the legal points at issue. As he put it, "The Supreme Court has on a number of occasions expounded on the meaning of the clause, and the pronouncements are clear."[5] By arguing from such established legal standards the judge could create an appearance of impartiality and fairness by asserting that his decision was consistent with the evolving yet stable framework of codified law.[6]

The plaintiffs' case included three major proofs aimed at showing how Act 590 violated the Establishment Clause. Each proof engaged one of the three specific tests for applying the clause.[7] First, a proof was constructed that showed Act 590 did not have a secular legislative purpose. The plaintiffs did this by exposing sectarian motives behind legislative efforts to enact the bill. The judge agreed: "The only inference which can be drawn from these circumstances is that the Act was passed with the specific purpose by the General Assembly of advancing religion. The Act therefore fails the first prong of the three-pronged test, that of secular legislative purpose."[8] Second, the plaintiffs offered proof to demonstrate, as Overton would conclude, that "since creation science is not science . . . the *only* real effect of Act 590 is the advancement of religion."[9] Here the plaintiffs sought to show that creationists were not "real" scientists, and that creationism did not possess any scientific significance whatsoever. Act 590 thus failed the second criterion of constitutionality. Finally, the plaintiffs demonstrated to Judge Overton's satisfaction that Act 590 would create "an excessive and prohibited entanglement [of government] with religion" because compliance would involve federal and state authorities in "delicate religious judgments" when called upon to screen course materials and classroom discussions for religious references.[10]

Judge Overton decided that these three major contentions were convincing. In what follows, I shall focus primarily on the construction of the second proof.

The *McLean* decision makes clear that settling the controversy was not a simple matter of applying clearly defined legal standards to the facts of the case. There were important nonlegal, substantive issues to be settled before meaningful application of the law even could be possible. The judge had to seek out and apply nonlegal standards of reasonableness in adjudicating nonlegal issues—issues which, if left unresolved, could block meaningful applications of the law. A strategic rhetoric about science weighed heavily in shaping Judge Overton's nonlegal criteria. Specifically, he had to turn to the testimony of expert witnesses to locate criteria for distinguishing thought and action as "scientific" or otherwise.

McLean v. Arkansas illustrates how rhetoric is made about science when rhetors deal with exigences that I shall call demarcation exigences. In such exigences the central ambiguities to be addressed have to do with setting boundaries that separate science from nonscience.

In this case scientific insiders and their supporters sought to modify the demarcation exigence in ways that would include specifically scientific professional services and "exclude providers of 'similar' services who falsely claim (according to insiders) to be within the profession."[11] In contrast, creationists and their supporters tried, but failed, to make convincing claims about the scientific standing of creationists and creationism. I shall focus on the plaintiff's rhetorical strategies in attempting to define creationists as scientific outsiders and creationism as beyond the pale of scientific reasonableness.

Eric Holtzman and David Klasfeld have said that "the essential thrust of the plaintiffs' effort was to demonstrate that, at its heart, creationism is not science at all but rather is thinly disguised religion."[12] I believe Holtzman and Klasfeld are right about the overarching thrust of the plaintiffs' case. I also believe, however, that criteria were needed for demarcating who was and who was not a member of "the recognized scientific community." Two questions would seem very important to Judge Overton in this regard: Who is a scientist? and What is science?[13] The plaintiffs answered with two rhetorical tactics: by (1) arguing or implying that creationists are not real scientists because they do not display qualities of professional character that a Mertonian vision prescribes as appropriate for a scientific ethos; and (2) taking conjectural stands against creationism on technical issues in order to show that creationists' postulates did not constitute reasonable scientific claim-making.

The rhetorical exigence that needed solution involved ambiguities about what it means to think and to act as scientists. In this kind of rhetorical situation rhetors could not assume that the rhetorical audience, in this case Judge Overton, would simply treat scientific ethos as something given. They had to recognize that perceptions of scientific ethos and of who possesses its qualities had to be constructed. In this case the plaintiffs tried to create rhetorically a Mertonian vision of scientific ethos. They therefore constrained Overton's perceptions of creationists as people who lacked the scientific virtues of communality, skepticism, universality, and disinterestedness. For these reasons he was persuaded that creationists did not think or act like scientists and, therefore, were not credible expert witnesses. Gieryn, Bevins, and Zehr explain:

> Attribution of these selected characteristics to science was effective for excluding creation-science from "professional" science: creationists testifying for the defense of Act 590 lack the credentials that certify success in passing by the gatekeepers at the boundaries of "professional" science [they lack communality]; they are dogmatic rather than skeptical; they refuse to admit the validity of universally obvious scientific facts; their conclusions are biased by a greater interest in advancing religion than in discovering truth [they are not disinterested].[14]

Expert witnesses for the plaintiffs sought to convince Judge Overton by arguing from the *topos* of communality. The general strategy was to show that creationists did not have good standing in the scientific community since they did not participate actively in its intellectual life. Specifically, the plaintiffs argued: (1) creationists did not publish frequently in authorized scientific journals; (2) creationists did not submit technical claims about creationism for peer review; and (3) creationists were not employed by leading scientific research institutions.[15] The judge found these and other arguments collectively convincing.

Creationists did not display the scientific virtue of communality because they failed to contribute to the production and dissemination of credible scientific knowledge, practices which characterize the raison d'etre for scientific communities. Gieryn et al. cite the example of a creationist geologist who admitted under cross-examination that "he had only two articles in standard scientific journals since getting his Ph.D. in 1955"—hardly a publication record for a geologist actively engaged in the intellectual life of his community.[16] Such admissions worked to diminish the scientific credibility of creationist witnesses for the defense.

A particularly persuasive argument was that creationists displayed a lack of scientific communality by failing to submit their technical claims about creationism to peer review. Creationists had argued that they could not publish in standard journals due to various forms of editorial bias. They argued that the scientific community, specifically evolutionary biologists, were guilty of "censorship," "country-club exclusion," and of trying to keep theories that "were incompatible with their personal or philosophical views 'out of the marketplace of ideas.'"[17] However, no proponent of creationism offered an instance of a rejected submission to a scientific journal. Overton's opinion shows he concluded that the absence of reputably published articles supporting creation science indicated that creationists were either unwilling to submit to or unable to meet the standards recognized by scientific journals. In his opinion, Overton directly defended the scientific community and its members as appropriate agencies for deciding what does and does not count as reasonable science:

> Creation science, as defined in Section 4(a), not only fails to follow the canons defining scientific theory, it also fails to fit the more general description of "what scientists think" and "what scientists do." The scientific community consists of individuals and groups, nationally and internationally, who work independently in such varied fields as biology, paleontology, geology and astronomy. Their work is published and subject to review and testing by their peers. The journals for publication are both numerous and varied. There is, however, not one recognized scientific journal which has published an article expressing the creation science model described in Section 4(a). Some of the State's witnesses suggested that the scientific community was "close-minded" on the subject of creationism and that explained the lack of acceptance of the creation science arguments. Yet no witness produced a scientific article for which publication had been refused. Perhaps some members of the scientific community are resistant to new ideas. It is, however, inconceivable that such a loose knit group of independent thinkers in all the varied fields of science could, or would, so effectively censor new scientific thought.[18]

The "scientific" claims of creationists had not made a ripple in the scientific community. That was enough to shatter their scientific ethos with the judge. Most creationist literature was published by Creation-Life Publishers. This publisher was not accepted as providing a legitimate scientific forum for testing ideas; it was, and is, recognized as a religiously motivated publishing company.

Plaintiffs and their witnesses also claimed creationists were not to be found holding positions at leading research institutions, once again arguing from the *topos* of communality. Harold Morowitz, a biophysicist at Yale, said he could not name a creationist who held a position in an Ivy League university. He added, "I can't give you the name of an Ivy League school, graduate school, or journal which houses a flat earth theorist either."[19]

Such testimony, coupled with the fact that scientists testifying for the plaintiffs did have sterling credentials, further diminished the credibility of the defendants' allegedly scientific witnesses.

The scientific credibility of creationists was also attacked on grounds that they did not have the skeptical inclinations of "real" scientists.[20] They were depicted as dogmatically asserting the inerrancy of Scripture. It was further pointed out that most were members of the Creation Research Society, a leading creationist organization. That organization required of voting members the equivalent of a master's degree in a scientific or technical area, but witnesses for the plaintiffs argued that the CRS was not a true association of scientists because members were required to take oaths of fidelity to beliefs in God and in the literal truth of Scripture.[21] These arguments impressed Overton, who included the CRS's official statement in his opinion:

> (1) The Bible is the written Word of God, and because we believe it to be inspired thruout (sic), all of its assertions are historically and scientifically true in all of the original autographs. To the student of nature, this means that the account of origins in Genesis is a factual presentation of simple historical truths. (2) All basic types of living things, including man, were made by direct creative acts of God during Creation Week as described in Genesis. Whatever biological changes have occurred since Creation have accomplished only changes within the original created kinds. (3) The great Flood described in Genesis, commonly referred to as the Noachian Deluge, was an historical event, worldwide in its extent and effect. (4) Finally, we are an organization of Christian men of science, who accept Jesus Christ as our Lord and Savior. The account of the special creation of Adam and Eve as one man and one woman, and their subsequent Fall into sin, is the basis for our belief in the necessity of a Savior for all mankind. Therefore, salvation can come only thru (sic) accepting Jesus Christ as our Savior.[22]

In the view of experts and the judge who chose to rely on their testimony, signing this statement demonstrated that creationists did not possess the skeptical outlook that is proper for scientists.

The scientific ethos of creationists was diminished further by their persistent refusal to share in scientific consensus about noncontroversial technical claims.[23] To use Merton's term, they did not display the quality of universality, which involves willingness to grant the status of truth, at least provisionally, to claims tested against impersonal criteria and thus certified as consonant with observations and with consensually embraced opinions about scientific knowledge. Even some of the defense's own expert witnesses gave testimony consistent with the charge. For instance, Chandra Wickramasinghe was chosen to give testimony for the defense. Wickramasinghe was known for a controversial theory of origins which, as Overton put it, "has not received general acceptance within the scientific community."[24] The plaintiffs chose to exploit Wickramasinghe's disagreements with creationism rather than to center attention on his maverick status in relation to the mainstream scientific community.[25]

That Overton was persuaded by this strategy is evident from the fact that in his opinion he claimed Wickramasinghe "demonstrated" the falsity of creationisms' two-model approach to the origins of life, and cited his statement that "no rational scientist" would explain the earth's geology with constructs that postulate an earth that is less than one million years old or a single worldwide flood.[26] Overton seems to have considered it a telling point that even maverick scientists shared in the professional consensus that the two-model approach to origins and the geological postulates of creationism constitute unreasonable scientific claims.

Creationists' scientific and legal ethos was also challenged on the ground that, as proponents of Act 590, they lacked both the scientific and legal quality of disinterestedness. Their scientific rhetoric was not motivated by knowledge-related aims,[27] nor was their legal rhetoric pursuing secular objectives; both served to conceal an objective that could not be sought openly with scientifically or legally reasonable rhetoric—to teach religious doctrine in the public schools. These rhetorical masks were removed during the trial. The defense had argued that what ought to count in Establishment Clause cases was the letter of the act and not the intentions of its authors, sponsors, and defenders. That argument did not persuade Overton, who said in his opinion that assessment of legislative purposes was appropriate in this case.[28] After hearing all of the arguments, he concluded that Act 590 "was simply and purely an effort to introduce the Biblical version of creation into the public school curricula."[29]

The plaintiffs' arguments and evidence convinced Overton that the creationists lacked disinterestedness both in science and in sponsoring Act 590. In his opinion, he cited a deposition from Paul Ellwanger, the act's author. The judge first implied that Ellwanger lacked proper legal and scientific credentials for advocating the act: he was "a respiratory therapist who is trained in neither law nor science."[30] Overton then quoted from a letter put in evidence that showed Ellwanger had religious rather than scientific objectives: "I view this whole battle as one between God and anti-God forces, though I know there are a large number of evolutionists who believe in God." The letter said further:

> If you have a clear choice between having grassroots leaders of this statewide bill promotion effort to be ministerial or non-ministerial, be sure to opt for the non-ministerial. It does the bill effort no good to have ministers out there in the public forum and the adversary will surely pick at this point. . . . Ministerial persons can accomplish a tremendous amount of work from behind the scenes, encouraging their congregations to take the organizational and P.R. initiatives. And they can lead their churches in storming heaven with prayers for help against so tenacious an adversary.[31]

Overton summed up Ellwanger's "ultimate purpose" for Act 590 with a quotation from another letter: "the idea of killing evolution instead of playing these debating games we've been playing for nigh over a decade already."[32] These excerpts are but a sample of documentary evidence used by the plaintiffs to show that creationists and their supporters sought religious aims through exercising political power; creationists could not exhibit an impressive scientific ethos because they were not involved in a disinterested search for scientific knowledge.

As I have illustrated, the plaintiffs argued from the *topoi* of communality, universality, skepticism, and disinterestedness. Arguments from these *topoi* persuaded the judge to deny the creationists' credibility as either scientists or disinterested supporters of Act 590. The use of these *topoi* was especially wise because the plaintiffs were thus able to focus directly on whether creationists possessed scientific ethos. Once the judge accepted that the controversy hinged on the scientific status of creationists, creationists were effectively dismissed as legitimate participants in the legal debate. The general point to be drawn from this instance is that these headings can apply logically in both scientific and nonscientific rhetorical situations. They identify a set of commonly accepted criteria for judging what is and what is not scientific conduct.

In his carefully and sharply worded decision the judge issued the forthright claim that the scientific community is the final, authoritative arbiter of rhetorical reasonableness on matters scientific. That is a widely but not universally held view in Western democratic societies. Once Overton adopted the descriptive definition of science as what is "accepted by the scientific community" and "what scientists do," he explicitly declared political standards of reasonableness irrelevant in deciding what is scientific. He asserted that "in a free society, knowledge does not require the imprimatur of legislation in order to become science."[33] Mark Herlihy, a lawyer for the plaintiffs, is right to observe that Overton's decision was significant because it allowed specialized communities to decide nonlegal issues concerning science and education.[34] Overton's selection of the scientific community as the source of grounds for reasonableness also denied creationists the moral, political, and religious standards that are their strongest sources of persuasive arguments with nonexpert audiences. Even so, conceptions of "the scientific community" and its "real" members vary and are not simply givens; they are arguable notions rhetorically constructed in specific problematic situations. Overton had to decide who was expert on scientific matters as a prelude to settling the legal issues before him. This was crucial because he himself lacked authority to decide matters of scientific reasonableness. Overton was persuaded to accept Mertonian standards for distinguishing "real" from "pseudo" scientists and, with those standards, he decided which testimony was "expert" regarding science and its profession in the public school curricula.

It is important to recognize the rhetorical implications of this point. Mertonian standards are very much at issue among scholars of scientific activity.[35] Overton was induced to make the judgment that Mertonian standards are realistic portrayals of scientific conduct and therefore ought to prevail in this case.

When rhetors construct scientific ethos, they must gauge the situational relevance of qualities they choose to promote as exemplifying "real" scientific character. In *McLean* the plaintiffs had an important logical advantage insofar as Mertonian *topoi* characterize conventional ways of thinking about scientific qualities of thought and conduct. They could draw upon the persuasive powers of presumption when invoking and applying Mertonian themes as constituting scientific ethos. Creationists and their legal defenders had to take on the burden of proving to the judge that either (1) Mertonian standards could be applied reasonably and favorably in assessing the scientific virtuosity of creationists' thought and conduct, or (2) there were sufficiently strong grounds for adopting alternative criteria for evaluating creationists' scientific ethos in this case. Failure to meet such burdens (and the plaintiffs' success at maintaining power or presumption) illustrates, once again, that constructing and evaluating scientific ethos is not arbitrary or happenstantial, but must meet criteria governing the rhetorical logic of science.

As I have indicated, in the rhetoric constituting his opinion Overton made clear that he thought intentions counted in interpreting the constitutionality of Act 590. He strengthened this claim by saying additionally that even if one relied only on the language of the act, one would still see "that both the purpose and effect of Act 590 is the advancement of religion in the public schools."[36] A sufficient sample of that language can be seen in section 4 of the act. I quote the section in full because its content was the source of arguments about the rhetorical reasonableness of creationism *as science:*

> (a) "Creation-science" means the scientific evidences for creation and inferences from those scientific evidences. Creation-science includes the scientific evidences and related inferences that indicate: (1) Sudden creation of the universe, energy, and life from nothing; (2) The insufficiency of mutation and natural selection in bringing about development of all living kinds from a single organism; (3) Changes only within fixed limits of originally created kinds of plants and animals; (4) Separate ancestry for man and apes; (5) Explanation of the earth's geology by catastrophism, including the occurrence of a worldwide flood; and (6) A relatively recent inception of the earth and living kinds.
>
> (b) "Evolution-science" means the scientific evidences for evolution and inferences from those scientific evidences. Evolution-science includes the scientific evidences and related inferences that indicate: (1) Emergence by naturalistic processes of the universe from disordered matter and emergence of life from nonlife; (2) The sufficiency of mutation and natural selection in bringing about development of present living kinds from simple earlier kinds; (3) Emergence by mutation and natural selection of present living kinds from simple earlier kinds; (4) Emergence of man from a common ancestor with apes; (5) Explanation of the earth's geology and the evolutionary sequence by uniformitarianism; and (6) An inception several billion years ago of the earth and somewhat later of life.
>
> (c) "Public schools" mean public secondary and elementary schools.[37]

The plaintiffs sought to undermine the scientific case for creationism not only through discrediting the ethos of its proponents, but also by showing that on technical grounds the above postulates and other claims do not constitute reasonable scientific claim-making. The plaintiffs had to argue in accordance with standards governing the rhetorical logic of scientific claim-making; but they also had to clarify those criteria and apply them at a level of generality sufficient to ensure their rhetoric would be situationally reasonable, given the adjudicating inexpert audience.

Among the criteria that Judge Overton eventually applied to evaluate the scientific status of creationism were those that differentiated scientific from nonscientific claims on the grounds that the former is (1) guided by natural law and (2) explainable by reference to natural law. These were, and are, arguable criteria. There are no universally acceptable demarcation criteria. For instance, philosopher of science Larry Laudan concluded that in *McLean* "the pro-science forces [were] defending a philosophy of science which is, in its way, every bit as outmoded as the 'science' of the creationists."[38] Laudan argued that the two demarcation criteria are far too restrictive to work effectively.[39] Nonetheless, Michael Ruse, Professor of History and Philosophy at the University of Guelph, helped through his testimony to establish the criteria Overton used in assessing the scientific merits of creationism.[40]

These differences of opinion justify three observations about rhetoric of and about science. First, Ruse and other witnesses established demarcation criteria in Overton's mind and did it *rhetorically*. In the process they deflected attention away from alternative demarcating standards. By rhetorical processes Overton was persuaded to accept as scientific and final the criteria Ruse offered. Second, it is likely that Overton was persuaded because, in addition to being authoritatively supported in the trial, the criteria allow extensive commonsensical rhetoric in their favor. Explaining phenomena in a lawlike manner can be readily expounded as a practical standard for scientific purposing. Third, once Ruse and others persuaded Overton to embrace the criteria, the way was opened for the plaintiffs to bring before Overton evidence that creationists not only failed to pursue such objectives but actually sought to subvert them. Joel Cracraft, Professor of Anatomy at the University of Illinois Medical Center and science adviser to the American Civil Liberties Union in the McLean case, would later make the point that science, in the creationist view, is "no longer . . . characterized solely by naturalistic explanations of natural phenomena—the creationists place supernatural explanations within the legitimate domain of science."[41] Even if at some time creationists' postulates were found to be true, it would remain the case that their methods of argument in *McLean* and the ethos they brought to the trial simply could not accord with the criteria governing "reasonable" science that were established in Overton's mind during the trial.

There has been some disagreement about whether the plaintiffs framed the points at issue appropriately in the McLean case. Laudan contended that the best course would have been to center discussion on issues involving comparison of creationism and evolutionism against available evidence. In his view creationists make claims that "are testable, they have been tested, and they have failed those tests."[42] Instead of framing the issues as demarcation problems, Laudan thought the plaintiffs could have made a stronger case by focusing on the comparative technical merits of creation and evolution:

> The core issue is not whether Creationism satisfies some undemanding and highly controversial definitions of what is scientific; the real question is whether the existing evidence provides stronger arguments for evolutionary theory than for Creationism. Once that question is settled, we will know what belongs in the classroom and what does not. Debating the scientific status of Creationism (especially when "science" is construed in such an unfortunate manner) is a red herring that diverts attention away from the issues that should concern us.[43]

In rhetorical terms Laudan was advising critics of creationism to focus on translative and qualitative issues. His recommendations have logical merit, but there were and are rhetorical reasons for critics of creationism not to follow his advice. One reason is that creationists are especially adept at arguing translative and qualitative issues involving comparisons of creationism with evolution before general audiences. The other reason is that if legal issues are to be met, it is necessary that creationism be denied the status of even a *weak* science. Acts like the Arkansas one cannot otherwise be judged unconstitutional.

Creationists have their greatest rhetorical successes when they are able to frame issues in much the same way as Laudan recommended. Typically, creationists frame issues *translatively* about *interpretive* exigences and *qualitatively* about *evaluative* problems. They seek to control debate by invoking central questions like: Does creation science or evolution science provide the more meaningful explanation of origins? Does creation science or evolution science possess greater significance or value?" Once these questions are implanted in the minds of an adjudicating but largely scientifically illiterate laity, creationists are able to assume the scientific legitimacy of creationism, implying that comparative judgments are being made about the scientific merits of both theories.[44] Typically, creationists argue that evolutionary theory cannot be applied meaningfully to a field in which phenomena are not fully known and understood.[45] They then imply that they are responding to translative issues about interpretive exigences, and they treat alleged difficulties in applying evolutionary constructs to problems such as origins as positive evidence for creationist postulates. A similar strategy is used to engage qualitative issues regarding the comparative value of evolutionary and creationist postulates. Alleged interpretive difficulties with evolutionary constructs are seized upon and used to imply that creationist postulates have greater empirical accuracy or consistency. These strategies beg the question of whether creationism is science, although creationists can adopt the style of scientists engaged in comparative debate. In *McLean* the plaintiffs sought to disarm these strategies by demanding positive claims to support the scientific reasonableness of creationism.[46] The crucial *stasis* point of the case was conjectural: "What is *scientific* about creationism?"

The Establishment Clause of the First Amendment bars the teaching of religion. As Ruse said, the plaintiff's best strategy was the one they adopted: showing "that creation-science is less than a weak or bad science. It is not science at all."[47] In the event, Ruse was right. Judge Overton decided to look to the scientific community for standards of scientific reasonableness, and he adopted the Mertonian framework for deciding which witnesses properly spoke for that community. The plaintiff's emphasis on what is and is not science was the right strategy given Overton's analysis of the critical *stasis* in the case.

The stands taken by the plaintiffs against creationists illustrate standard and quite predictable rhetorical strategies that scientific insiders are likely to use when trying to dismiss the claims of perceived outsiders on technical grounds. By establishing that creationism is *not* science, the plaintiffs could exclude creationists from the universe of scientific discourse, and the argument placed the burden of asserting positive scientific proofs on the creationists. The plaintiffs argued that the postulates of creationism did not contribute to solving problems that concern scientists as scientists. Their practices confirm the theoretical formulation I offered earlier for conjectural *stases* in science. The plaintiffs did in fact raise and address conjectural subissues concerning evidential, interpretive, methodological, and evaluative problems. The specific questions were: Is there scientific evidence for creationism's postulates? Are there scientifically meaningful constructs within the theoretical framework of creationism? Do creationists use reliable scientific procedures and techniques when pursuing their investigations? Do creationists' claims have scientific significance? The plaintiffs adduced evidence to support negative answers to all four subissues.

In their turn the defense cited evidence they believed demonstrated the explanatory power of creationist postulates. This was again turning to translative and interpretive issues without having settled the conjectural issue: But is creationism *science,* so that it belongs in a *science* curriculum? We know from Overton's decision that he decided the conjectural issue was the real issue in the case, and, of course, the defendants had evaded that "Is it *science*?" stand. In Overton's view the creationists' translative/interpretive proofs of creationism's superiority over evolutionary theory were irrelevant to that question. For example, Overton said:

> In efforts to establish "evidence" in support of creation science, the defendants relied upon the same false premise as the two-model approach contained in Section 4, i.e., all evidence which criticized evolutionary theory was proof in support of creation science. For example, the defendants established that the mathematical probability of a chance chemical combination resulting in life from non-life is so remote that such an occurrence is almost beyond imagination. Those mathematical facts, the defendants argue, are scientific evidences that life was the product of a creator. While the statistical figures may be impressive evidence against the theory of chance chemical combinations as an explanation of origins, it requires a leap of faith to interpret those figures so as to support a complex doctrine which includes a sudden creation from nothing, a worldwide flood, separate ancestry of man and apes, and a young earth.[48]

It would be insufficient to analyze Overton's reasoning here with principles of formal logic only. He was clearly deciding what the focus of logical argument ought to be. He was considering what the issue, not the form of argument, should be. The plaintiffs' focus on "Is creationism *science?*" was the critical question for Overton. The defendants' focus on "Which scientific formulations are better?" begged that question. This is shown by how Overton decided each conjectural subissue concerning problems of evidence, interpretation, methodology, and evaluation.

The judge pointed to lack of scientific evidence when dismissing the claims of creationists: "The arguments asserted by creationists are not based upon new scientific evidence or laboratory data which has been ignored by the scientific community."[49] With this statement the judge dispatched the creationists' complaints of having been unfairly ignored by the scientific community. He further indicated he was not convinced by creationists' arguments against the empirical adequacy of evolutionary theory. For instance, creationists argued that Robert Gentry's discovery of radioactive polonium haloes in granite and coalified wood constituted evidence against the empirical adequacy of evolutionary theory. The judge dismissed Gentry's discovery as "a minor mystery which will eventually be explained."[50] In any case, it did not constitute reliable evidence for creationism.

The plaintiffs addressed conjectural issues about interpretation by focusing on the langauge of the act as evidence of its religious motivation and nature. Testimony focused on the language of creationism postulates. Theologians helped to establish the "inescapable religiosity" of the language.[51] Langdon Gilkey and other theologians testified that notions such as "creation from nothing" and "creator" identified themes peculiar to Western religious thought. All scientific witnesses argued that the term "kinds" lacked any fixed scientific meaning.[52] Overton was apparently influenced by this line of argument, for he said the ideas of a sudden creation from nothing, worldwide flood, and relatively recent inception were all too clearly found in the Old Testament to hold scientific meaning.[53] He also accepted the notion that "kinds" was too imprecise to serve as a scientific construct.[54]

Plaintiffs further undermined the scientific case for creationism by questioning creationists' ability to make scientifically meaningful interpretive statements about received tenets of evolutionary theory. For example, section 4 (b) (2) of the act stated that mutation and natural selection explained how the existence of present life forms derived from earlier forms. Ayala and Gould testified that this was contrary to biologists' beliefs that other processes also play evolutionary roles. Overton was sufficiently impressed to recount their testimony in support of his conclusion that section 4(b) of Act 590 was "simply a hodgepodge of limited assertions."[55]

The plaintiffs raised interpretive problems in creation science by using rhetorical strategies that left the creationists literally without any terministic means for reasonable scientific expression. The plaintiffs situated all terms and constructs of creationism within an all-encompassing framework of Fundamentalist Christian thought. Through these conjectural challenges plaintiffs leveled from a variety of directions an overarching charge against creationism as science: all meaningful "scientific" explanations broached by creationists are ultimately derived from the belief that God has made supernatural interventions. Accordingly, Overton was moved to write: "If the unifying idea of supernatural creation by God is removed from Section 4, the remaining parts of the section explain nothing and are meaningless assertions."[56]

In *McLean v. Arkansas,* as in other forums, creationists were unable to persuade a scientifically oriented (and legally oriented) audience because they lacked legitimate theory and empirical evidence from which to argue for the scientific reasonableness of, say, explanations, taxonomies, and predictions consonant with their theory of origins. We have here an instance in which arguments cannot be developed from promising *topoi* because the rhetors do not have the content those arguments would require. The seriousness of this rhetorical limitation was exacerbated by the fact (as brought out in the trial) that creationists seldom engaged in controlled, empirical investigations when doing "creation research." Ruse pointed out that creation scientists "do little or nothing by way of genuine test." He contended "experimental or observational work" is almost totally absent from creationist literature, observing that their argumentation "proceeds by showing evolution (specifically Darwinism) wrong, rather than by showing Creationism right."[57] Ruse's testimony appears to have impressed the judge, who directly referred to creationists' methodology as "indicative that their work is not science."[58] There was little with which creationists could counter such claims, hence they could not meet the plaintiffs' conjectural, methodological challenge.

As I have said, the plaintiffs in *McLean* sought to close the universe of scientific discussion to creationists. Cumulatively their rhetoric worked toward creating a conjectural *stasis* in evaluation, raising the ultimate evaluative question: Does creation science have any scientific value at all? In the trial, as elsewhere, those opposing creation science challenged its value most strongly with specific arguments drawn from the *topoi* of empirical accuracy, internal consistency, and external consistency.[59] Creationists have few specific arguments they can draw from those *topoi.* Perhaps this is why in *McLean* they tried to shift the evaluative points of stand to the qualitative (Which of the two theories of origins possesses stronger values, scientific or otherwise?) and interpretive points at issue to the translative (Which of the two theories provides the more meaningful constructs?)—arguing refutatively by centering attention on alleged problems in evolutionary theory while averting myriad problems inherent in their own. In the McLean case, of course, not even those lines of argument were fruitful before a judge who adopted the Mertonian tests for assessing scientific ethos and who insisted the crucial issue of the trial was, Is creationism something that ought to be in a *science* curriculum?

A fundamental principle of the rhetorical logic governing scientific discourse is that reasonable claim-making involves maintaining or expanding a scientific community's comprehension of natural order and doing so empirically by following prescribed scientific methods for investigating technical exigences. Attacks on creationists' scientific ethos worked toward establishing the conclusion that they not only violate but actually subvert this rhetorical standard for scientifically reasonable purposing and claim-making. Conjectural stands exposed creationists' inability to make scientifically reasonable, positive claims about the evidential grounds, theoretical constructs, scientific methods, and special scientific values of so-called creation science. Consequently they could not satisfy the minimal tests of reasonable scientific discussion; they were left quite literally without *any* scientific matter for discussion. Once the judge decided the point for decision was conjectural and that evolutionary theory was not on trial, the outcome of the case became predictable.

Creationists failed to make rhetorically effective discourse for this case. One should recognize, however, that this judgment says nothing about the ultimate truth, or ethicality, or rationality of either side's discourse. Rhetoricians have argued over whether rendering these kinds of judgment is a sufficient ambition for rhetorical criticism.[60] I do not choose to enter into those speculations here, but I do point out that the analysis and evaluation of *rhetorical invention* that I have been offering does let us see the kinds of inventional choices that work and that fail where the key question is: Are we dealing with *science* here? Under the court's criteria, creationism was demarcated from science. This does not change the fact that the claims of creationists have been authorized as reasonable by members of many audiences. This fact simply reflects the reality that different intellectual communities use different arrays of logical tests for what is reasonable. The very appeals and arguments that failed to convince established scientists and Judge Overton remain good reasons for a good many political and religious audiences. For example, the creationists made much of the mathematical improbability that chance chemical combinations could create life from nonlife. If, in some other rhetorical setting, the "truth" of evolutionary theory were the point at issue, this could become a genuinely persuasive logical argument. Detailed study of creationist rhetoric as, say, political rhetoric rather than as rhetoric of science would probably show there are many more opportunities for winning favorable judgments than emerge from a study of creationist rhetoric as rhetoric of science.

THE DOUBLE HELIX AND THE RHETORIC OF SCIENCE

Watson and Crick's announcement in *Nature* of their double-helical structure for DNA (Deoxyribose Nucleic Acid) was an event that is also suitable for critical analysis using the special theory of scientific rhetoric I have offered. The announcement provides a useful contrast to discourse related to *McLean v. Arkansas*. Through it we have an opportunity to test the theory of how the rhetoric of science works, using an instance of what most regard as a virtuoso scientific presentation. *McLean v. Arkansas* provides instances of rhetoric about science that was made for a technically unskilled audience, whereas Watson and Crick's article is an example of rhetoric that induced expert audiences to cooperate in thought and practice with the symbolic orientation the rhetoric provided.

Despite its technical details the DNA controversy is accessible to laypersons. Comprehensive historical accounts are available for general readers.[61] The DNA story is also interesting because it presents a picture of a demystified science. Participants in the controversies associated with discovery of DNA's structure displayed an all-too-human side of science that is gossipy, envious, vain, secretive, and sometimes mean-spirited[62]—a side that is often concealed when scientists make rhetoric of and about science. The rhetoric associated with the Watson-Crick announcement was productive of new knowledge and continues to be admired; yet, once more, the governing logic of that rhetoric was situationally determined and was evaluated against topical criteria built up within the special culture of scientific communities.

The straightforward presentation of Watson and Crick's historic announcement, "A Structure for Deoxyribose Nucleic Acid," could easily mislead a reader.[63] One could conclude that discovery of the structure was a rather noncontroversial affair. In fact, there was a large degree of conflict and competition in the search for a way of resolving the technical exigence of DNA's molecular structure. Francis Crick and James Watson were not alone when striving to solve this problem at Cavendish Laboratory at Cambridge. Linus Pauling was also working on the problem at the California Institute of Technology, and prior to Watson and Crick's announcement he had published an unsatisfactory solution to the problem.[64] Maurice Wilkins and Rosalind Franklin were working on the problem at King's College in London. A number of other prominent scientists played lesser parts in the DNA drama,[65] but the five scientists I have named were the main rhetors who sought to solve convincingly the exigence of the molecular structure of DNA.

The straightforward reportage characteristic of published literature about DNA's structure also obscures the fact that solving the technical exigence required addressing the rhetorical exigence of convincing members of a scientific community that one had found a situationally reasonable way of solving or modifying the structural problem. The articles themselves result from strategic choices made according to criteria governing the informal logic of scientific rhetoric in view of situational contingencies. Rhetorical purposing was (as it always is) one such critical choice.

Rhetorical purposing in science must make clear that what is being addressed are problems in a scientific community's comprehension of natural order. Watson and Crick had the rhetorical advantage that the technical exigence concerning the molecular structure of DNA was already established in the minds of many scientists. Their concern, therefore, would involve how best to frame the problem rhetorically when offering their bihelical model as a solution.

Much was known about the DNA molecule when Watson and Crick started working earnestly to solve its structure. There was growing awareness that DNA was the hereditary material, a point intimated by Oswald Avery's work (with co-workers MacLeod and McCarty) as early as 1944 and proclaimed more boldly by A. D. Hershey and Martha Chase in 1952. The chemical composition of DNA had been known since the 1930s. Two purine (adenine and guanine) and two pyrimidine (thymine and cytosine) bases were attached to sugar and phosphate groups to form nucleotides. A "backbone" or "chain" was formed when sugar groups (with attached bases) were chemically linked with phosphate groups. One or more of these chains ran the length of the molecule. X-ray data supplied important information about the molecule's dimensions: it consisted of substructures that repeated every 3.4 angstrom units, the entire structure repeated every 34 angstrom units, and the molecule's diameter was a constant 20 angstroms. For Watson and Crick the problem to be settled was how all this and other information came together in a structural configuration that could also account for the molecule's biological function.

Watson and Crick framed the technical exigence as an *interpretive* one involving ambiguity in the precise meaning of the molecule's structure.

They were not forced to frame the exigence in this way, but given their methodological constraints and their overt desire to assure claims to priority for the discovery, this was the wisest choice. That there were alternative ways of framing the problem is illustrated in the papers published along with the double-helix announcement.[66] For instance, Franklin and Gosling reported the results of x-ray diffraction studies. Their conclusions were consonant with Watson and Crick's proposal, but they framed the problem primarily as *evidential,* involving structural ambiguities which, in their view, could be clarified only through obtaining reliable experimental evidence. However, Crick and Watson focused primarily on offering a theoretically sound solution to the problem, buttressed by experimental data supplied by others' research.

In offering the bihelical proposal as a solution, Crick and Watson made a reasonable case by engaging conjectural, definitional, and qualitative issues involving interpretation. They engaged each of those issues and in proper sequence. Further, they drew lines of thought from problem-solving and evaluative *topoi* as they rendered their claims reasonable and fitting responses to conjectural, definitional, and qualitative points at issue. Watson and Crick's announcement is thus prototypical of reasonable rhetoric in science which responds fittingly to a significant technical exigence.[67]

The opening sentence of the article indicates that the central purpose is to modify an important technical exigence: "We wish to suggest a structure for the salt of deoxyribose nucleic acid (D.N.A.)." This sentence directs readers' attention to the technical problem of DNA's molecular structure and intimates the authors possess a solution. The interpretive nature of this problem and the theoretical nature of their solution become clearer in the following two paragraphs, where the two authors argue that, at present, there are no meaningful models available that solve the existing problem in ways that are consonant with received biochemical principles and established structural data.

In their second and third paragraphs Watson and Crick implicitly invoked and explicitly addressed *conjectural stases* about interpretation, and thus they opened a gateway in received knowledge for their proposed solution. The issue they addressed was: Is there or is there not a reliable model for interpreting recalcitrant structural data about DNA? They answered the question in the negative by examining two failed models. First they dealt with the Pauling and Corey model:

> A structure for nucleic acid has already been proposed by Pauling and Corey. . . . Their model consists of three inter-

> twined chains, with the phosphates near the fibre axis, and the bases on the outside. In our opinion, this structure is unsatisfactory for two reasons: (1) We believe that the material which gives the x-ray diagrams is the salt, not the free acid. Without the acidic hydrogen atoms it is not clear what forces would hold the structure together, especially as the negatively charged phosphates near the axis will repel each other. (2) Some of the van der Waals distances appear to be too small.

Watson and Crick's scientific rhetoric presented two reasons for rejecting this model: The hydrogen bonds allegedly holding the triple helix together "made chemical nonsense" (to use Judson's language) because the structure did not contain necessary binding forces, and the atoms were packed so tightly at the core of the structure that they violated distances at which atoms would get in each other's way.[68] These reasons persuade without need for supporting calculation because, as Bazerman pointed out, they are grounded in received biochemical knowledge. Bazerman identified the logical structure warranting both refutative claims: "If a model does not match existing theory which is believed to accurately describe nature, then the model must be dismissed."[69]

Of the second available model Watson and Crick said in their third paragraph,

> Another three-chain structure has also been suggested by Fraser. . . . In his model the phosphates are on the outside and the bases are on the inside, linked together by hydrogen bonds. This structure as described is rather ill-defined, and for this reason we shall not comment on it.

Bazerman summed up the likely rhetorical impact of this paragraph: It was not worth looking at Fraser's model in more detail because it was too ambiguous to allow meaningful discussion "against the framework of codified knowledge or against measurable aspects of the object."[70]

Watson claims he and Crick did not plan even to mention Fraser's work until Wilkins and Franklin insisted they do so after having read a draft of their article. Wilkins and Franklin thought acknowledging Fraser (a member of their lab) was warranted because he "had considered hydrogen-bonded bases prior to our [Watson and Crick's] work." Watson writes that he and Crick thought Fraser should be excluded because "his idea did not seem worth resurrecting only to be quickly buried." In their view Fraser's previous work did consider hydrogen-bonded bases—an important component of the eventual solution—but did so wrongly: he had "dealt with groups of three bases, hydrogen-bonded in the middle, many of which we now knew to be in the wrong tautomeric forms."[71] Thus, the third paragraph seems a succinct instance of a situationally negotiated compromise regarding priority of important knowledge claims.

In sum, Watson and Crick dismissed both models as scientifically unreasonable because they were not consistent with received chemical knowledge. With these models disposed of, Watson and Crick were ready to show how the bihelical model could fill the void in knowledge.

The overarching exigence Watson and Crick now addressed was: What is the precise structural arrangement of the DNA molecule? This question raised crucial subissues: Is the structure helical? Most thought so, though Franklin suspected the A-form of DNA was not helical. If helical, how many helices were there? Most believed there had to be more than one. Pauling and Corey's and Fraser's failed models contained three, and so did an aborted effort by Watson and Crick. What were the positions of the sugar-phosphate "backbones" and attached bases in the structure? Pauling and Corey thought the backbones were on the inside with the bases outside, and at one time so did Watson and Crick; Franklin thought the backbones were on the outside with the bases on the inside. Even if the backbones were on the outside and the bases inside, there was still the complex question: How are the chains related in the molecule? These were questions of configuration. The scientific exigence was also a *rhetorical* exigence because its resolution involved use of suasory discourse to induce authorization for whatever resolution was proposed.[72]

Watson and Crick sought to fill the void in comprehension of the DNA structure by engaging a *definitional* issue about the interpretive meaning of their proposed model. They were answering the central question: What is the model for the molecular structure of DNA? As they described the bihelical model, they proposed how some subissues should be settled. The two scientists started to describe their model in their fourth paragraph, making amplifying reference to a "purely diagrammatic figure on the left margin":

> We wish to put forward a radically different structure for the salt of deoxyribose nucleic acid. This structure has two helical chains each coiled round the same axis (see diagram). We have made the usual chemical assumptions, namely, that each chain consists of phosphate diester groups joining B-D-deoxyribofuranose residues with 3′, 5′ linkages. The two chains (but not their bases) are related by a dyad perpendicular to the fibre axis. Both chains follow right-handed helices, but owing

> to the dyad the sequences of the atoms in the two chains run in opposite directions. Each chain loosely resembles Furberg's model No. 1; that is, the bases are on the inside and the phosphates are on the outside. The configuration of the sugar and the atoms near it is close to Furberg's "standard configuration," the sugar being roughly perpendicular to the attached base.

The paragraph continues with additional information about the molecule's dimensions: the diameter is 20 angstroms, the height between stacked parallel bases (running up the middle of the two chains) is 3.4 angstroms, the height of one complete turn of each helix is 34 angstroms, and there was a 36° rotation from one nucleotide to the next on each of the backbones.

This description of the double helix sought to answer three questions and partially address a fourth. Is the structure helical? The answer given was yes. How many helices? The answer was two. Where are the sugar-phosphate backbones and the bases placed in the structure? The answer: backbones on the outside, bases on the inside. How are the chains related? Part of the answer was that for the molecule to repeat itself at the height of 34 angstroms both backbones would have to twist around the outside of the core a complete 360° (i.e., 36° rotation from one nucleotide to the next for a total of ten nucleotides) and in *opposite* directions (with the sequence of atoms in one matching the other in reverse). The dyadic symmetry of the molecule alluded to in paragraph four accounted for the spatial arrangement of the backbones,[73] but readers were not yet told how the backbones were tied together at the bases.

The description rendered the double-helical model reasonable by indicating ways in which its structural features fit within the framework of existing knowledge. The two scientists made "the usual chemical assumptions." They also said that the placement of the bases and the sugar-phosphate backbones "loosely resembles Furberg's model," and that the configuration of sugar and atoms was "close to Furberg's 'standard configuration.' "[74] Not so incidentally, the controlling dimensions reported for the molecule would be confirmed by x-ray data contained in the experimental papers published along with their article in that issue of *Nature*. All of these statements indicated points at which the model fit what was already known about the DNA molecule. The authors still had to explain how the two backbones were joined together.

One view was that the DNA molecule was built up from like-with-like base pairs. According to this view the two strands would run in the same direction, with adenine paired with adenine, thymine with thymine, guanine with guanine, and cytosine with cytosine. Watson and Crick's great insight was that they saw the strands joined together in a new relation: the strands ran in opposite directions with the hydrogen-bonded base-pairing mechanism joining the strands in a complementary fashion, with adenine joining thymine, and guanine connecting with cytosine. Watson and Crick explained how the two helices are hydrogen-bonded together at the bases in paragraphs six, seven, and eight:

> The novel feature of the structure is the manner in which the two chains are held together by the purine and pyramidine bases. The planes of the bases are perpendicular to the fibre axis. They are joined together in pairs, a single base from one chain being hydrogen-bonded to a single base from the other chain, so that the two lie side by side with identical z-coordinates. One of the pair must be a purine and the other a pyrimidine for bonding to occur. The hydrogen bonds are made as follows: purine position 1 to pyrimidine position 1; purine position 6 to pyrimidine position 6.
>
> If it is assumed that the bases only occur in the structure in the most plausible tautomeric forms . . . it is found that only specific pairs of bases can bond together. These pairs are: adenine (purine) with thymine (pyrimidine), and guanine (purine) with cytosine (pyrimidine).
>
> In other words, if an adenine forms one member of a pair, on either chain, then on these assumptions the other member must be thymine; similarly for guanine and cytosine. The sequence of bases on a single chain does not appear to be restricted in any way. However, if only specific pairs of bases can be found, it follows that if the sequence of bases on one chain is given, then the sequence on the other chain is automatically determined.

Watson and Crick were clarifying the meaning of their bihelical structure for DNA, and while doing so they were solving the interpretive exigence that confronted scientists working in the area. Their model contained two sugar-phosphate backbones that wind along the outside of the fiber axis with the hydrogen-bonded complementary bases forming the structure's core. In his *Double Helix,* Watson tells readers they should think of the DNA structure as analogous to a "spiral staircase," with the base pairs forming the steps.[75] In their article the authors showed that they next to sought to complete their interpretive case by addressing qualitative issues concerning empirical applications of the model.

To make a complete response to the question of the molecular structure of DNA they must now address the *qualitative* issue: Which interpretive applications of the bihelical model are more meaningful? To maximize possibilities of scientists' judging the model to be reasonable, Watson and Crick had to argue this issue by establishing the empirical grounds for the model.

The model was a proposal about the physical world, and had not yet been confirmed through empirical investigation. Nonetheless, what the proposed model claimed to do was represent an actual empirical structure. Its success would depend ultimately on the ability to induce ontological commitments—the belief that DNA's molecular structure is not *like* a double-helix but *is* a double helix. Bazerman identifies a fundamental logical principle governing the reasonableness of ontological claim-making in science: "The claim of representing an actual structure rather than creating an approximate model results in a strong requirement for correspondence between data and claim."[76] Watson and Crick sought to meet this standard by establishing closeness of fit between theoretical and available data claims. Specifically, they took a stand on the qualitative issue of empirical application by arguing that their model possessed empirical adequacy and explanatory power.[77]

Following the description of the base-pairing mechanism, in paragraph nine Watson and Crick showed the closeness of fit between their model and the well-established Chargaff ratios: "It has been found experimentally that the ratio of the amounts of adenine to thymine, and the ratio of guanine to cytosine, are always very close to unity for deoxyribose nucleic acid." This paragraph offered at one stroke two reasons for support of the model. The reader could comprehend that the model was empirically adequate insofar as it fit the Chargaff ratios. Moreover, the model actually explained the data. Chargaff had established that in all DNA the molecule-to-molecule ratios of adenine to thymine and of guanine to cytosine were not far from unity. When juxtaposed with the base-pairing mechanism that holds the model together, the Chargaff ratios were viewed as inevitable consequences of the mechanism. Elsewhere Crick recalled his experience during a meeting with Chargaff in which he had learned about the ratios: "I suddenly realized, by God, if you have complementary replication, you can *expect* to get one-to-one ratios."[78] Chargaff himself would later write that the model was "the most plausible inference from the base-pairing regularities earlier discovered by us in many DNA preparations."[79] One would expect readers of the article in *Nature* to be affected in similar fashion. As a result, readers would tacitly supply one of the premises by which they would become persuaded of the model's explanatory power.

Watson and Crick continued to address the qualitative issue by clarifying where meaningful applications of the bihelical structure would seem unlikely. They wrote in paragraph ten that "it is probably impossible to build this structure with a ribose sugar in place of the deoxyribose, as the extra oxygen atom would make too close a van der Waals contact." (The bihelical model would probably not apply to RNA because the atoms would be too close together, given the distance at which atoms get in each other's way.) Unless there was evidence to suggest that the alleged van der Waals contact was wrong, readers could agree that it was reasonable not to apply the model to RNA, given received chemical principles and available evidence. In contrast, Judson claimed the Pauling and Corey paper appeared to make the "biochemically bizarre" claim that their model was correct for both RNA and DNA.[80] Thus, one rhetorical problem with that paper was its failure to engage satisfactorily the qualitative issue about their model's range of potentially reasonable applications.

Watson and Crick made their most persuasive argument for empirical adequacy in paragraph eleven. They again addressed a qualitative issue about interpretation. The first sentence of the paragraph read, "The previously published x-ray data on deoxyribose nucleic acid are insufficient for a rigorous test of our structure." They referred to a 1947 study by Astbury and to a more recent publication (1953) by Wilkins and Randall. Watson and Crick were unable to claim experimental precision using these data, but they did suggest that their model was empirically adequate because, "so far as we can tell, it is roughly compatible with the experimental data, but it must be regarded as unproved until it has been checked against more exact results." Further on in the same paragraph the authors strengthened their claims to empirical adequacy by pointing to the companion papers in the same issue of *Nature:* "Some of these [more exact results] are given in the following communications. We were not aware of the details of the results presented there when we devised our structure, which rests mainly though not entirely on published experimental data and stereochemical arguments."[81]

Immediately following Watson and Crick's article appeared one by Wilkins, Stokes, and Wilson entitled "Molecular Structure of Deoxypentose Nucleic Acid." Next came an article by Franklin and Gosling, "Molecular Configuration in Sodium Thymonucleate." From a rhetorical perspective the articles enhanced the persuasive prospects of Watson and Crick's potentially adequate proposal.

The Wilkins et al. article reported x-ray evidence from their studies of intact sperm heads and bacteriophage that gave patterns suggesting DNA in living creatures had a helical structure much like Watson and Crick's model.[82] The Franklin and Gosling paper presented a crucial diffraction photograph of the DNA fibre in its B-form, and asserted their interpretation of x-ray evidence was consonant with Watson and Crick's proposed structure.[83] Taken together, these articles added to the persuasiveness of Watson and Crick's argument for empirical adequacy.[84]

It was Wilkins who suggested that he (and his co-workers) and Franklin and Gosling be allowed to publish their data simultaneously with Watson and Crick's article.[85] That this was done enabled Watson and Crick to enhance their model's immediate persuasiveness. They could now at least strengthen their claims to empirical adequacy beyond referring to previously published structural data. Indeed, their use of the two experimental papers created the impression that strengthening data were, to some degree, independently confirming the structure while conserving claims for Watson and Crick's originality of thought in constructing the model. Consider the following carefully crafted statements (my italics) in paragraph eleven: "We *were not aware of the details* of the results presented there [in the two articles] when we devised our structure, which rests mainly *though not entirely* on published experimental data and stereochemical arguments." And in the fourteenth and final paragraph: Watson and Crick had been "stimulated by a knowledge of the *general nature* of unpublished experimental results and ideas of Dr. M. H. F. Wilkins, Dr. R. E. Franklin and their co-workers at King's College, London."[86]

The form of the claim to empirical adequacy and the manner of crediting other investigators reflect the facts that: (1) empirical adequacy is an important logical structure for scientific claim-making; and (2) after their article had been given its basic form, Watson and Crick continued to adjust it rhetorically, as the circumstances of their rhetorical situation changed.

It is worth mentioning that Watson and Crick obtained important x-ray information while constructing their model, but the way in which it was secured became a matter of some rhetorical controversy about whether they and others acted responsibly as scientists. Watson and Crick had benefited greatly from learning of Franklin's unpublished work. Wilkins showed Watson Franklin's picture of DNA in the B-form, which would eventually appear in her article with Gosling. It seemed to Watson to confirm information about the helical structure of the molecule.[87] Later, Max Perutz gave Watson and Crick a copy of Franklin's unpublished research report submitted to the Biophysics Committee of the Medical Research Council. This report contained useful information about the A-form, the role of water in the transition of DNA from the A- to the B-form, and related data.[88] Watson and Crick had all of this useful information while working on their structure. That Franklin was never asked permission to transmit this information to Watson and Crick gave rise to questions about ethics.[89]

With the history of the twelfth paragraph now available to us, we again perceive the authors' deliberation about rhetorical strategy as they composed a scientific report. In that paragraph Watson and Crick chose to address the important conjectural issue about evaluation: Is the DNA double-helical model scientifically significant? (This issue was anticipated since the second sentence of the first paragraph reads: "This structure has novel features which are of considerable biological significance.") We now have evidence that the twelfth paragraph as actually published reflects a conscious inventional choice about what it was situationally reasonable to claim for the model's significance. An almost cryptic sentence appears in the paragraph: "It has not escaped our notice that the specific pairing we have postulated immediately suggests a possible copying mechanism for the genetic material." According to Watson, Crick originally wanted to expand the discussion of the genetic implications, but finally "saw the point to a short remark."[90] According to Crick, Watson did not wish to make the claim about replication (let alone extend discussion) because he feared the structure might prove to be empirically inaccurate.

We are, in fact, confronted with a paragraph in which what ultimately proved to be a very important point was made with exceptional brevity because the two authors had to come up with a verbal compromise that was the product of their differing notions of what they could actually claim and of what the scientific community would accept. Crick explains how the compromise came about:

> This [paragraph twelve] has been described as "coy," a word that few would normally associate with either of the authors, at least in their scientific work. In fact it was a compromise, reflecting a difference of opinion. I was keen that the paper should discuss the genetic implications. Watson was against it. He suffered from periodic fears that the structure might be wrong and that he had made an ass of himself. I yielded to his

> point of view but insisted that something be put in the paper, otherwise someone else would certainly write to make the suggestion, assuming that we had been too blind to see it. In short, it was a claim to priority.[91]

When we have the after-the-fact accounts from Watson and Crick, we can see that their rhetorical difficulties were of two completely intertwined types: (1) what they, as a team, took as sufficiently reasonable to present publicly in their rhetorical situation, and (2) what the scientific community reading the article would think of their ethos as scientists. One could hardly have a clearer instance of how inseparable scientific knowledge and the logic of rhetorical adaptation can become when scientists make discourse.

Within a few weeks after publication of their seminal article Watson and Crick composed a more speculative article elaborating on the genetic implications of their model.[92] The pair decided to present their claims more boldly when they realized how strongly the x-ray evidence contained in the King's College papers supported their proposed structure. As Crick put it, "The main reason was that when we sent the first draft of our initial paper to King's College we had not yet seen how strongly the x-ray evidence supported our structure." He concluded, "We were . . . delighted to see how well their evidence supported our idea. Thus emboldened, Watson was easily persuaded that we should write a second paper."[93]

Watson and Crick did not strongly emphasize the *fruitfulness* of their model. They did touch this *topos* when they mentioned the genetic copying mechanism in paragraph twelve. They also suggested immediate, potential fruitfulness in the next paragraph when they promised to publish later "full details of the structure, including the conditions assumed in building it," and "a set of co-ordinates for the atoms." Those data would help make the discovery more useful to others. Nothing else in the article directly asserted their model's fruitfulness. The presence of the companion articles in the same issue of *Nature* would also suggest that there was much important theoretical and experimental work to be done, because Watson and Crick's model connected to important features of the other articles. As events turned out, the fruitfulness rhetorically implied in that 1953 issue of *Nature* became a reality. A spate of theoretical and experimental research followed the DNA double helix into publication.[94] As a basis for answering questions, the double helix spurred great developments in molecular biology and continues to serve as an exemplary historical paradigm for that field.

Failure to develop the *topos* of fruitfulness might seem an oversight, but more importantly from my vantage point, such failure resulted from a *rhetorical* decision. What if Watson had let Crick discuss the genetic implications of the model, as Crick wanted? The model's fruitfulness would then become a major feature of the article's persuasiveness. However, there would also be greater latitude for readers to assess how strongly (or weakly) the *topos* applied to specific claims made. Fruitfulness was not developed more fully because of a disagreement about rhetorical strategy, but the compromise that resulted in the twelfth paragraph was rhetorically wise for at least three reasons. The brevity of the statement (1) established their claim to priority for the replication mechanism; (2) indicated the model's immediate fruitfulness by suggesting the structural solution also identified the molecule's biological function—replicating genetic material; and (3) allowed them to avoid complex and potentially controversial problems concerning precisely how the chains separate and unwind during replication or how the base sequences act like a genetic code.[95] Elaborate discussion was not needed to ensure perceptions of fruitfulness in this rhetorical situation. Indeed, given the prominence of the technical exigence, the brief statement would readily invoke conceptions of fruitfulness in the minds of others.

In his insightful rhetorical criticism of the DNA announcement, Halloran observes how the description of the model's features, by drawing enthymematically upon scientists' appreciation for *elegance,* constituted a persuasive argument for the scientific value of the bihelical structure.[96] Chargaff succumbed to the rhetorical appeal of elegance when he called the model an "aesthetically pleasing solution."[97] The descriptive paragraphs four, five, six, seven, and eight show the remarkable symmetry and balance of the DNA double helix. This is excellently exemplified by the complementary base-pairing mechanism: when one base is adenine, the other must be thymine; if one base is guanine, the other is cytosine. Lest the reader think I attribute too much to the implicit persuasive force of elegance, let me briefly identify some of the explicit statements made by Watson and Crick about their model. In *The Double Helix,* Watson wrote, "A structure this pretty just had to exist."[98] In an essay in *Nature* commemorating the twenty-first anniversary of their first paper, Crick denied the claim of some commentators that it was the "style" of the two scientists that contributed to the forceful impact of their discovery, making this revealing statement:

> Rather than believe that Watson and Crick made the DNA structure, I would rather stress that the structure made Watson

> and Crick. After all, I was almost totally unknown at the time and Watson was regarded, in most circles, as too bright to be really sound. But what I think is overlooked in such arguments is the *intrinsic beauty of the DNA double helix*. It is the *molecule which has style*, quite as much as the scientists.[99]

Examining the rhetorical dimensions of Watson and Crick's presentation still more thoroughly makes it clear that neither style nor elegance can alone explain the article's success, although both qualities doubtless contributed. At least equally significant, I contend, was the wisdom with which the authors chose their points to address and the persuasiveness and logical relevance of the *topoi* from which they drew arguments.

Watson and Crick did what scientists must do when addressing their peers. They selected a rhetorical purpose aimed at addressing and modifying a central exigence before the scientific community: What is the molecular structure of DNA? They addressed this question by framing the central rhetorical exigence as interpretive, and by sequentially engaging conjectural, definitional, and qualitative *stases* that otherwise could have blocked full appreciation for their solution to the problem. Moreover, Watson and Crick chose wise and logical *topoi* as grounds from which to argue and render their claims reasonable for other scientists. Their article in *Nature* was persuasive because by habit, reason, strategy, or all three, the authors argued that their model for DNA promised *empirical adequacy* and possessed qualities of *consistency, elegance, explanatory power,* and *immediate fruitfulness.*

My claim is that these are among the standard logical tests that are applied by scientists (and others) to argumentation in scientific discourse. By developing these lines of thought, Watson and Crick were astute rhetors of science. They chose wisely from among the available means of logical persuasion as they sought to induce authorization of the bihelical model as a reasonable and significant expansion of scientific comprehension of natural order. My rhetorical analysis has shown that the influence of the article was not simply the result of its bold ideas. Astute rhetorical decisions entered in, several times very consciously. Watson and Crick displayed consummate rhetorical skill: it was a virtuoso performance in scientific rhetoric with power to compel an authorizing audience's immediate, appreciative understanding, if not their immediate full conviction.

Scientists could, and did, make inventional choices other than those made by Watson and Crick. This is characteristic of making rhetoric about science. There are always alternative inducing possibilities. It is also the case that scientists are constrained by their chosen methodological orientations when assessing the comparative situational reasonableness of these possibilities and making inventional choices from among them.

The quest for the DNA structure raised the question of what method of inquiry was most useful at the time. Two methods had gained prominence as scientifically reasonable ways of solving molecular structures. Franklin and Wilkins attempted to solve the structure with x-ray crystallography. Simply put, this method involved passing x-ray beams through a crystal of the substance under investigation so that when atoms and molecules deflect the beams a diffraction pattern is produced on a photographic plate placed beyond the crystal. Intensive analysis of this pattern is then made to deduce the three-dimensional shape of the structures causing it. And this method was successful. By 1953 crystallography was solving structures of ever-increasing complexity. The method had been used earlier to solve molecular structures of metals and minerals, and since then was continually undergoing technical refinements while being extended to more complex, biologically important molecules.[100] Watson and Crick favored the more speculative but nevertheless rigorous method of chemical model building. Pauling had developed this method, and used it to discover the alpha helix, a molecular structure found in proteins. Watson and Crick learned from his example and, in similar fashion, sought to infer DNA's three-dimensional structure from manipulating Tinker Toy-like physical models designed to obey the principles and rules of structural chemistry.[101]

Both methods were recognized and respected approaches to solving molecular structures, but favoring one over the other would necessarily invoke orientations that variably constrain what constitutes scientifically reasonable framing and solving of structural problems. Judson shows how these two methodological approaches toward investigating molecular structures, together with the theoretical interest in solving the molecule's genetic function, generated the central scientific orientations governing both investigative and rhetorical reasonableness in the problem-solving situation:

> Seen as an unusual essay in pure crystallography applied to a fibrous structure, as Franklin wanted to solve it, the problem was static. Exactly where was each group of atoms located and how was it aligned? Seen as model-building, by Watson or Pauling, the correct shapes of the parts and at least the broad limitations on the whole had to be sorted out from the work of biochemists and crystallographers, and the puzzle was geometrical: How could the bits all be assembled without breaking any of the rules? Seen as functioning biology, and perhaps

> Crick understood it this way most clearly, the question became dynamic: How did the structure dictate the assembly of replicas of itself—on itself, from itself—so accurately that DNA could carry the hereditary message? These were all really versions of the same question. The answer to any would be the answer to all.[102]

X-ray crystallographers were not against modeling, but their methodological orientation led them to wish to assemble experimental facts before trying to construct models. In dealing with DNA they had good reasons. There had been some glaring failures due to premature molecular modeling. I have already mentioned the failure of Pauling and Corey's model. Near the end of 1951 Watson and Crick advanced a model which mistakenly put the backbones in the center of the structure, a mistake later clarified by Franklin's fine crystallographic work.[103] Like most crystallographers, Franklin wanted to avoid such pitfalls.

Because of Franklin's methods of investigation she had communicative problems different from those faced by the more speculative Watson and Crick. Her methodological orientation influenced how she chose her rhetorical purposes, framed central issues, and selected from available scientific *topoi* when trying to make and present scientifically reasonable claims.

Franklin's methodological orientation prescribed that problems being investigated should be treated as *evidential* problems. For instance, consider her struggle to solve the DNA molecule's structure in its A-form. She believed one had to clear up ambiguities concerning the availability, reliability, and meaning of experimental evidence before constructing and presenting a scientifically reasonable solution to that problem. However, her x-ray diffraction investigations led her to evidence in startling conflict with data she had about DNA in the B-form. Investigatively, but not yet rhetorically, she had hit upon a *conjectural* obstacle in evidence. She interpreted the photograph as having intensities too asymmetrical to fit the helical structure in the A-form. She made rhetoric about the matter when she spoke to Crick, seeking to engage him in rhetorical discussion about a conjectural issue of evidence: "Are there x-ray data supporting a helical structure for DNA's A-form?" She argued that her evidence did not indicate a helical structure, and implied that the general helical hypothesis for DNA's molecular structure was threatened with possible experimental disconfirmation.[104]

Franklin argued from her measurements (i.e., from the *topos* of quantitative precision) that she had discovered an evidentiary flaw in the helical view, implying that she might have found a significant anomaly in the helical hypothesis. Watson and Crick were not convinced this anomaly was significant, and they simply denied the accuracy of Franklin's data. They never saw her photograph, but they nevertheless argued that her measurements must be wrong. Crick, recollecting the encounter, reveals that he and Watson patronized Franklin; rather than offering refutative arguments they issued simple denials, raising conjectural problems by questioning the reliability of her evidence: "When she told us DNA couldn't be a helix, we said, 'Nonsense.' And when she said but her measurements showed that it couldn't, we said, 'Well, they're wrong.' You see, that was our sort of attitude."[105] As events turned out, Watson and Crick were correct, even though this certainly was not due to careful assessment of Franklin's claim. Nevertheless, Franklin's insistence that she had discovered a conjectural problem with the helical view exemplified what Crick called being misled by the data, which, in his view, is less likely to occur when data are used heuristically. In contrast, Franklin especially valued experimental data, "hard facts," and for her, modeling was the heuristic endeavor.

Franklin was a good experimentalist, and she realized that to reject the helical hypothesis persuasively, at least in the A-form, she would first need additional evidence to remove the conjectural obstacle. Only one analyzed photograph so far yielded conflicting data. If she could secure similar data, she could argue more persuasively from the *topos* of experimental corroboration. This evidence never materialized. She took pictures almost every day from August 14 until September 9, and again on October 14, in an unsuccessful search for results that would corroborate her data and make a persuasive empirical argument.[106] She never could take a convincing conjectural stand about evidence against those who advocated the helical structure as a plausible hypothesis; indeed, she herself would later publish positive evidence for a helical structure for DNA in the A-form.[107]

Franklin would not have expressed her problem in the language of rhetorical theory, but from that point of view and according to Judson's accounts of her efforts, she was being frustrated by conjectural and definitional obstacles. Given her methodological orientation, she could not offer a general hypothesis about DNA structure until she could locate empirical data that allowed a consistent interpretation. Even if she found the needed evidence against helical structures, she would still face the definitional problem of determining and explaining persuasively what the evidence meant. She tried several analogical lines of reasoning—perhaps the molecule was a parallel "double-sheet structure" of "banana-like units" or shaped like the "figure-8"—but in the end she had to reject these solutions.[108]

Franklin's struggle was not simply a matter of laboratory work; it was a struggle to *express* in a clear and defensible form a hypothesis about DNA structure. Judson reports on Franklin's integrated investigational-rhetorical activities:

> "If there is a *flat* banana-like unit in structure, with banana axis parallel to fibre axis," she now wrote, the explanation must be "rather a double-sheet structure"—and she sketched a dozen more tiny bananas viewed from the top and then from the side. The red notebook that had begun as her bench record of x-ray camera work was now the place to which she retreated in the attempt to put her difficult gropings into coherent form. That Wednesday, she fretted about helices, bananas, and double sheets for a couple of pages. By the following Monday the calculations had been extended, and her notes were concerned with peaks that she would expect to observe but that were not to be found. On the fifteenth, in five curt lines, she concluded that "there is no narrow straight chain of high density parallel to the axis" of the fibres. Then not a line more for five weeks.[109]

Franklin eventually stopped pondering the structure of the A-form and shifted her attention back to her excellent photographs of the B-form. Clearly she could not remove the definitional obstacle about the meaning of A-form DNA data.

At one point in her quest for the DNA structure Franklin had almost all parts of the puzzle calculated, including the diameter of the molecule, outside placement of backbones, and recognition that both A- and B-forms must be helical, with two helices per molecule. According to Aaron Klug, Franklin was two steps from finding the double helix: "What she missed was the presence of the dyad—the symmetry of the molecule. Which Francis [Crick] had seen immediately from her information about the space group. And even though her notes earlier showed that she knew about Chargaff's ratios, she never got to the base pairing."[110] Franklin came close, but from a rhetorical perspective her central accomplishment was to supply Watson and Crick with evidentiary means of persuasion for rendering their proposed model reasonable, scientifically.

As with all empirically minded investigators, x-ray crystallographers are likely to pursue fairly predictable strategies when making rhetorical presentation of findings. They will direct attention to *evidential* exigences about what does and what does not exist, engaging conjectural, definitional, qualitative, and sometimes translative *stases* when trying to solve problems of this kind. Consider the *stases* addressed by Franklin and Gosling in their article published along with the double-helix announcement. By supplying and analyzing a reproduction of an x-ray diagram of structure B, these two scientists provided an affirmative answer to the *conjectural* question: Is there reliable evidence for determining DNA's B-form structure? They began by asserting, "The x-ray diagram of structure B (see photograph) shows in striking manner the features characteristic of helical structures."[111] Franklin and Gosling assumed the helical hypothesis, claiming it was "immediately possible, from the x-ray diagram of structure B, to make certain deductions as to the nature and dimension of the helix."[112] They often shifted to the *definitional* question: What does the evidence mean? Drawing upon the *topos* of quantitative precision, they showed how important structural measurements and related details gleaned from analysis of the x-ray diagram fitted meaningfully within a general bihelical construct. All of this culminated in answering *qualitative* questions about the interpreted evidence: Which empirical judgments are warranted by the evidence? Having earlier admitted that x-ray data did not directly prove the helical structure, they now concluded: "While we do not attempt to offer a complete interpretation of the fibre-diagram of structure B, we may state the following conclusions. The structure is probably helical. The phosphate groups lie on the outside of the structural unit, on a helix of diameter about 20 A. The structural unit probably consists of two co-axial molecules."[113] Franklin and Gosling then extended their qualitative applications of interpreted evidence to the bihelical model: "Thus our general ideas are not inconsistent with the model proposed by Watson and Crick in the preceding communication."[114]

Franklin and Gosling, like Watson and Crick, argued conjectural, definitional, and qualitative *stases,* and in that sequence. They argued differently from Watson and Crick by framing rhetorical exigences evidentially rather than interpretively. In the final analysis we can discern an appeal to scientific significance, an effort to answer a conjectural question about evaluation: Why are these data and analyses significant? Franklin and Gosling tried to establish the significance of their data and accompanying analyses through tacit appeal to the promise of their empirical accuracy, a scientific value that stands atop any evaluative hierarchy that crystallographers are likely to embrace.

As important as data were, Watson and Crick's methodological orientation seemed to allow them greater flexibility in using experimental data than did Franklin's. This does not mean chemical model-building lacks rigor. Indeed, a scientifically significant model of a molecular structure must be constructed according to the most exacting standards of consistency with received chemical principles and will ultimately prove empirically accurate only when tested against x-ray evidence. But Watson and Crick displayed greater willingness than Franklin to use data heuristically, while she appeared to be driven by concerns about data. This reflects how different methodological orientations vary in how they emphasize the values all scientists share, thereby yielding differing conceptions about what constitutes scientifically reasonable claim-making. Franklin's methodological orientation magnified the value of empirical accuracy—a value embedded in all orientations insofar as they are legitimate *scientific* orientations—but did so in greater degree than did Watson and Crick. Crick's justification of using facts heuristically reveals that he and Watson worked from a methodological orientation which specially valued *simplicity*. Indeed, he implied x-ray crystallographers failed to find the alpha helix of protein (discovered by Pauling through chemical model-building) because their methodological orientation did not sufficiently value simplicity in scientific inquiry:

> You must remember, we were trying to solve it with the fewest possible assumptions. . . . There's a perfectly sound reason—it isn't just a matter of aesthetics or because we thought it was a nice game—why you should use the *minimum* of experimental data. The fact is, you remember, that we knew that Bragg and Kendrew and Perutz had been *misled* by experimental data. And therefore every bit of experimental evidence *we* had got at any one time we were prepared to throw *away*, because we said it may be misleading just the way that 5.1 reflection in alph keratin was misleading. . . . They missed the alpha helix because of that reflection! You see. And the fact that they didn't put the peptide bond in right. The point is that evidence can be unreliable, and therefore you should use as little of it as you can. And when we confront problems *today*, we're in exactly the same situation. We have three or four bits of data, we don't know which one is reliable, so we say, now, if we discard that one and assume it's wrong—even though we have no evidence that it's wrong—then we can look at the *rest* of the data and see if we can make sense of *that*. And that's what we do *all the time*. I mean, people don't realize that not only can

> data be wrong in science, it can be *misleading*. There isn't such a thing as a hard fact when you're trying to discover something. It's only afterwards that the facts become hard.[115]

At least one scientist was not impressed by how Watson and Crick generally approached science, regardless of the virtues that can be attributed to chemical model-building. Chargaff attacked Watson and Crick, charging that their conduct displayed qualities exemplifying a degenerate scientific ethos. Chargaff's contempt for what he perceived to be their arrogant and unthorough approach in investigating scientific problems shows through his assessment of their first two *Nature* articles:

> The tone was certainly unusual: somehow oracular and imperious, almost decalogous. Difficulties, such as the even now not well-understood manner of unwinding the huge bihelical structures under the conditions of the living cell, were brushed aside, in the Mr. Fix-it spirit that was later to become so evident in our scientific literature. . . . I could see that this was the dawn of something new: a sort of normative biology that commanded nature to behave in accordance with the models.[116]

However one may assess Watson and Crick's ethos as reflected in their approach to solving problems, it is nonetheless clear that throughout the search for DNA's structure a central methodological question was, Is molecular model-building or x-ray crystallography the better method for investigating the molecular structure of DNA at this time? Of course, the methods of investigation scientists chose allowed them to beg this question. Both methodological orientations as scientific orientations share scientific values, but each nonetheless can vary in claims about scientific significance by magnifying the situational relevance of some values over others. How scientists stood on this issue thus constrained how they applied the informal logic of scientific argument while they sought ways of making scientifically and situationally reasonable claims about the rhetorical exigence of clarifying DNA's molecular structure.

CONCLUSION

The two bodies of discourse I have examined in this chapter present rhetoric about science and rhetoric in the doing of science. I believe my analyses have shown (1) that rhetors choose from among alternative ways of saying what they have to say; (2) that the wisdom of their choices is constrained by the standards of reasonableness predominating in the situated audience addressed (one audience being both legal and scientific and the other a technically focused community of scientists); (3) that where doing science is the subject of discourse, a finite set of themes, values, and criteria constrains what it will be necessary and appropriate to say; (4) that these constraints on management of content are neither irrational nor formally logical but constitute the premises of a rhetorical logic that informs and regulates judgments of scientific discourse; and (5) that these constraints can be characterized as adaptations, for science, of principles of general rhetorical theory concerning ends, points at issue, and *topoi.* In this as in preceding chapters, I have observed what rhetors dealing with science do, and I have found that they do what general theory of rhetoric predicts they will do. They function, however, within the special framework of constraints and thematic options associated with "scientific method." The creativity that takes place within this framework abjures certain themes, values, and criteria of the reasonable that would be appropriate in other enterprises and settings. Such is the rhetorical logic and the inventional enterprise in rhetoric in and about science. It is a logic regulating choices among alternative but possible purposes, choices among alternative but possible issues, and choices among alternative but possible topics or patterns of discussing.

11

CONCLUSION

I began this inquiry with the hypothesis that inventional theory of rhetoric supplies a special way of thinking about situated, discursive human interactions, including those interactions peculiar to science. The general theory of rhetoric that has evolved from ancient times to the present posits how rhetoric persuades or, as I prefer to put it, how discourse is created to induce cooperative attitudes and actions. That body of theory purports to identify the compositional processes and decisions generally involved in producing such inducing discourse. I have argued that scientific discourse aims at inducing cooperative attitudes and actions from a particular kind of situated audience and therefore ought to be amenable to rhetorical description and analysis. To this end I have explained the general theory of how sound rhetoric is invented, have then adapted that theory specifically to science, and have presented examples of the kinds of analysis and interpretation the theory of scientific rhetorical invention allows.

It has long been generally agreed that doing science entails rhetorical styling, organizing, and presenting data and interpretations. I agree, but take the further view that the rhetoric of science is strategically *created* with a view to securing acceptance as reasonable by a special kind of audience. It is based on a particular kind of topical logic.

I have shown that composers of scientific discourse do, in principle, make the same general kinds of choices as other rhetors. When functioning as scientists, however, scientists' choices have to be guided by the special values and premises that define what is reasonable as *science.* These are not precisely the same values and premises that render political, theological, and other kinds of rhetoric reasonable. In the later chapters of this book I have undertaken to identify and illustrate the operation of these unique rhetorical constraints in a wide variety of scientific situations.

Like other rhetors scientists exercise reason, strategy, habit, or all three to do three major kinds of rhetorical choosing. First, they must choose rhetorical aims that at least appear to further their audiences' ultimate values. This involves pursuing maintenance or expansion of a scientific community's comprehension of natural order. Second, like other rhetors they must address exigences or ambiguities that concern their situated audiences. The special kinds of exigences that scientists confront prove to be evidential ambiguities about the empirical existence of phenomena, interpretive confusions and uncertainties about the meanings of constructs, evaluative ambiguities regarding the scientific significance of claims made, and methodological problems concerning appropriate scientific actions. To be "fitting," specific rhetorical responses must identify and address one or more of four issues arising from the problem area: conjectural, definitional, qualitative, and translative. Third, scientists, like other intellectual communities, have special ways of thinking and discoursing reasonably when deliberations are about scientific matters. Partly guided by the nature of their materials and by the nature of their claims, but partly guided also by what seems likely to persuade their situated audiences, scientific communicators choose from among certain standard topical lines of thought peculiar to scientific discussion. These topics identify the substance and the form of contents and structures thought to be logically appropriate in discussing subjects scientifically. The scientific discourses examined lead me to propose that, apart from ethos topics, scientific lines of development fall into three general classes: problem-solving, evaluative, and exemplary themes.

I devoted chapter 10 to detailed examination of efforts to produce successful scientific rhetoric in two kinds of rhetorical situations. The first was controlled by a demarcation exigence concerning whether or not creationism is science. That exercise in critical analysis shows that if we identify and evaluate the inventional choices made in relation to *McLean v. Arkansas,* we learn that the case for creationism failed as science because (1) creationists could not avail themselves of topics and structured arguments that would establish a credible scientific ethos for themselves; (2) both scientific and legal audiences would see their discourse as following an unscientific logic that subverts rather than contributes to expanding and maintaining the scientific community's comprehension of natural order; (3) creationists did not choose scientifically convincing stands against plaintiffs' challenges to the claim structure of creation science. Regarding this last point, plaintiffs demanded that creationists supply scientific evidence for claims, scientifically meaningful terminology and conceptions, evidence of scientific tests of claims, and scientifically convincing values for their purported science. But creationists did not satisfactorily address these conjectural issues. The plaintiffs pointed out these scientific flaws and went on to conserve and bolster their own positions in all four respects. In the trial orthodox scientists and their legal defenders made inventional choices amplifying the point that creationists could not qualify as doing science because what they chose to say and to leave unsaid violated the rhetorical value system that informs the thinking of all scientific communities in the West. In this instance of what I have called a "lay" rhetorical situation, orthodox scientists resolved the demarcation exigence in their favor, inducing the judge to decide that creationists and their activities fall outside the boundaries of reasonable science.

At other points in this book I have shown that similar kinds of rhetorical shortcomings account for allegations that other purporting scientists do not qualify as doing science or doing science well. However, it would be difficult to find a better exemplar of an unreasonable rhetoric of science than creationism. The kinds of shortcomings found in creationists' rhetoric during *McLean* are not always insurmountable when addressing inexpert audiences, but they are rhetorical shortcomings that will always lead rhetors to failure when addressing situated scientists as audiences. The reason is that these rhetorical shortcomings in invention constitute evasions or violations of the logical criteria that scientists impose on all who purport to do science.

Watson and Crick succeeded rhetorically because they did virtually all that is expected in reasonable scientific argument within a technical rhetorical situation. A recognized scientific exigence existed. They framed that exigence as an *interpretive* problem: If available theoretical understandings of DNA's structure were not revised, the ambiguity in comprehension of natural order would continue. They directed attention to *conjectural* issues concerning how to interpret evidence and principles of DNA's structural nature. They claimed that no adequate model of this structure existed. They addressed *definitional* issues concerning the precise meaning of their own proposed bihelical model. And they addressed *qualitative* issues by arguing for the empirical adequacy of their model: it fit existing knowledge and theory and had explanatory power that opened the way for fruitful further research. Throughout Watson and Crick's argumentation, the evaluative standards for doing science were respected and endorsed in ways that would maintain the authors' scientific ethos with their peers.

Watson and Crick's practices are prototypical of successful rhetorical argumentation in science. Illustrations of successful scientific rhetoric cited elsewhere in this book also support that view.

The case studies and illustrations I have presented underscore the fact that a specifically scientific logic is at work whenever scientists try to make claims seem reasonable *as science*. This is true whether they address rhetoric *about* science to the laity or address rhetoric *of* science to peers. Of course, situational adjustments have to be made in response to the differing constraints imposed by these quite different audiences. In consequence, rhetoric *about* science generally (1) emphasizes themes concerning scientific ethos, (2) minimizes development of technical premises dictated by scientific *topoi,* and (3) incorporates extrascientific concerns, themes, and materials to enhance interest and promote understanding by lay persons. Unless such adaptations are made, rhetoric about science will not seem reasonable in lay situations. Nonetheless, such rhetoric cannot disregard the rhetorical logic *of* science lest its scientific reasonableness suffer. In other words, the rhetorical logic of science, as I have outlined it in this book, is the conventional and at least loosely understood logic of science in the West. Disregarding scientific values and premises renders discourse something other than "scientific."

At their most general level my claims are: (1) there is a systematic way of thinking about how to induce cooperative attitudes and actions on any subject; (2) standard rhetorical theory identifies the general pattern of such thinking; (3) the pattern involves solving three intellectual problems: (a) choosing a situationally reasonable and pragmatic communicative purpose; (b) identifying what needs, defects, or exigences are recognized or can be created within the situation where communication is to occur; and (c) developing available lines of thought, indexed by *topoi,* into specific kinds of arguments that can potentially evoke an audience's cooperation in modifying logically crucial issues. This, I contend, is the elemental logic of thinking about what to say when one wants to influence the character of a rhetorical situation on any subject whatever. In other words, this is the nature of all rhetorical invention.

When one turns from general theory to ask what the pattern of rhetorical invention is for scientific communication, the postulates of general rhetorical theory become more precise. One cannot think about purposing in precisely the same way as one moves from theological to political to legal to scientific subjects and audiences. The interests, values, and logical requirements of each community prescribe and so limit what can be legitimately purposed, designate what will be the significant points at issue, and specify the range of appropriate lines of argument.

I have focused on how the interests and values of scientific communities modify options in purposing, in isolating salient issues or *stases,* and in developing lines of argument with a view to seeming scientifically reasonable. I have done this by examining what was done in prominent instances of successful and unsuccessful attempts at persuading scientists. The evidence indicates that the general theory of rhetorical invention applies where science is done through discourse. Scientists weigh with varying degrees of wisdom what discursive practices have potentialities for persuading situated audiences. They do this using a topical kind of logic acquired through training, habit, and/or strategic deliberation. They choose purposes wisely or unwisely; they select issues which they then address; and they select lines of thought on the bases of which they build specific arguments and claims. Whether or not they consciously realize that these are recurring and systematic parts of their procedures, their rhetoric shows they make these decisions.

The unsuccessful rhetorical efforts of creationists and the challenged efforts of parapsychologists and Francine Patterson are instances of at least partial failures in deciding how to create reasonable scientific rhetoric. Their difficulties in qualifying as "real" scientists are traceable in part to their failure to use convincingly the standard *topoi* of scientific discussion and evaluation. On the other hand, the rhetorical achievement of Watson and Crick stands as a prime instance of scientific rhetoric that fully meets the logical principles of rhetorical invention for scientific forums. Their article was a fitting response to a systematic *stasis* analysis of the interpretive exigence they were addressing. It is plain, I believe, that the commonplace "There is science and then there is rhetoric" is, at the very best, inaccurate in significant cases.

Offering an inventional theory of the rhetoric of science does more than merely identify science as involving rhetorical activity. According to Bokena, associating rhetoric with science raises a crucial question:

> What can the rhetorical discipline say about science processes that will illuminate these processes in ways which are distinct from, yet beneficial to, scholars in other humanistic and social-scientific disciplines, the public at large, and even science itself? Such a response will enhance and facilitate future development of the rhetorical understanding of science.[1]

There are several appropriate responses to Bokena's question. I have shown there *is* a rhetorical dimension to doing science. That does not imply science is rhetoric. It is to say that we shall not fully understand how science is "done" unless we understand its rhetorical aspects as well as its other aspects. The rhetoric of science is certain to be influenced by, but is not identical with, the logic of scientific method, the psychology of scientific inquiry, the sociology and history of scientific enterprises. Rhetoric of science needs to conform to the constraints these features of science impose, and rhetoric of science needs to make use of the many but finite creative options these features offer for effective communication. Creating this sort of rhetoric involves thinking, not just about science per se, but also about how best to negotiate ideas within human interactions among scientists. The latter is the *rhetorical* aspect of what scientists do and need to do when creating and developing claims about science. Successfully negotiating ideas in this way entails knowing and observing the informal but logical requirements that scientists, as scientists, apply in evaluating allegedly scientific communications.

Scientific discourse shows that when scientists compose communications for scientific audiences, they seek to be judged reasonable where the logic of science is rigorously applied. To succeed in such situations they must design discourse that will be perceived as identifying, modifying, or solving problems that bear on their scientific community's comprehension of natural order. Makers of scientific discourse need to present themselves as intending to modify some specific problem their fellow scientists perceive as pertinent to their better comprehension. Purposing in science is thus a special kind of rhetorical purposing. It must insert claims into the framework of the responding scientists' professional concerns and do so precisely. Scientists seek to modify four kinds of scientific problems or ambiguities: *evidential, interpretive, evaluative,* and *methodological.* All scientific problems involve points of *stasis*—points of "stoppage"—that must be settled, waived, or otherwise removed as obstacles if clear modification of the problem is to be attained. These points of stoppage occur concerning *conjectural* matters, *definitional* matters, *qualitative* matters, and *translative* matters. Analysis of scientific discursive practices shows the standard themes, or *topoi,* commonly developed in scientific rhetoric include the problem-solving themes of *experimental competence, experimental replication, corroboration, observational competence, experimental originality, explanatory power, predictive power, taxonomic power, quantitative precision, empirical adequacy, significant anomaly,* and *anomaly solution.* Scientific rhetoric also draws from evaluative *topoi* that reflect the special values of scientific communities. These include *accuracy, internal consistency, external consistency, scope, simplicity* or *parsimony, elegance,* and *fruitfulness.* In addition, scientific discourse develops exemplary themes when it uses *examples, analogies,* and *metaphors* to justify claims as scientifically reasonable. Finally, discourse of or about science often develops themes related to scientists' ethos. The relevant *topoi* include *universality, skepticism, disinterestedness,* and *communality.*

Such is the structure and character of the rhetorical logic underlying much, perhaps all, discursive presentations of scientific information.

I do not contend this theory and description of scientific rhetoric explains all that doing science amounts to. Neither have I exhausted the scope of a theory of rhetorical invention in science. The demands of that portion of a theory of scientific rhetoric that I have inferred from tradition and scientific discourse turn out to be a veritable checklist of topics normally taken up in teaching the basic principles of scientific method. This is not surprising, for those topics encompass the practical logic that governs doing science. The theory of the rhetoric of science which I have presented identifies and organizes these criteria and the patterns of logical judgment customarily used in creating and evaluating scientific discourse. Once we understand them and their functions as a special, topical logic, we can better understand the characteristics of successful scientific discourse and how and why some attempts at scientific discourse fall short of their makers' hopes. I believe there is a gain in this kind of understanding.

In recent struggles to discover and explain the "logic of science," positivistic principles of logic have come to be judged inadequate for that purpose. However, the further point is seldom if ever made that science, as a discipline, teaches a topical logic. My inquiry has highlighted the fact that it is that topical logic that provides the criteria by means of which scientific claiming is created and evaluated within scientific communities. Analyzing scientific discourse from a rhetorical perspective brings this reality to the fore and begins to answer the question: If we forgo positivism, what can we then say about how our claims are legitimized? Some have answered that scientific claims are legitimized by audiences or by consensus of a community. I believe this is to say too little. The special theory of rhetoric of science responds that claims are legitimized by being weighed against the topical logic that uniquely specifies the ways a claim may be rendered scientifically reasonable. Audiences of scientists do not, on the whole, derive their criteria for judgment from personal choice. They carry with them an organizable group of requirements that "the scientifically reasonable" must meet. Wise scientific rhetors will therefore consider those requirements in composing communications for their peers. The rhetorical theory for science itemizes and organizes those requirements and expectations.

Whether specific applications of criteria constituting the rhetorical logic of science are situationally reasonable is always a potential source of controversy and debate. The criteria do not apply uniformly from one situation to another. I have provided numerous instances of scientists disagreeing about which purposes should be pursued, which issues should be addressed, and which topical themes should be developed in making situationally and scientifically reasonable responses to technical exigences. Which criteria apply most reasonably to a given case is always a matter for situational negotiation and judgment.

Commentators on scientific activity can disagree about whether the conventional criteria governing the rhetorical logic of science could or should be followed in particular fields of human endeavor. Some contend that the rhetoric of science applies fittingly to problems in comprehending natural order but is inappropriately extended to social and political problems. Some critics of the social sciences would adopt this or a similar view.[2] Others call for entirely fresh conceptions of science to replace the criteria governing conventional scientific practices today. My intention has not been to engage these and other important critical issues directly. I do believe that regardless of how one stands on those issues, setting out the prevailing rhetorical logic of scientific argumentation is a necessary ingredient of any fruitful critical discussion of the nature and applications of science. Some benefits immediately follow from this rhetorical understanding of scientific activity.

What is the value of using and applying a theory of rhetorical invention in science? First, laypersons as well as scientists can understand more clearly a heretofore perplexing aspect of doing science. Doing science is not an exercise of formal logic or of following some neutral, procedural algorithm. Nor is it purely idiosyncratic or illogical. According to the theory I have offered, the logic of the rhetoric of science is a logic whose general pattern of reasoning takes this form: If the proposed claim is X and the situation has in it constraints A, C, D, and F, and the potential inventional options are P, Q, R, S, and T, which of these options can best render the claim reasonable in the situation offering these restraints and opportunities. The logic is exercise of situated deliberation about alternative possible ends, alternative possible issues, and alternative possible lines of argument; and it is carried out by adjusting criteria of scientific reasonableness to specific situational contingencies concerning the particular claim being made. The logic of rhetorical invention in science is not formal logic as that phrase is usually understood, but neither is it impulsive, haphazard, or otherwise nonlogical.

The theory I have offered points to the possibility of an entirely fresh conception of this kind of rhetorical logic. Such scholars as Chaim Perelman and L. Olbrechts-Tyteca, Ernesto Grassi, Walter Weimer, Walter Fisher, Leo Apostel, Stephen Toulmin, and many others have worked or are working on this problem.[3] My hope is that the theory of rhetorical invention in science I offer can forward that ongoing work.

The theory of invention for scientific discourse also serves to demystify what scientists actually do when they practice their craft. This is especially important in our age. We live in a world in which public issues must often be decided with the help of expert scientific testimony. In every other news story in modern days citizens are offered the testimony of scientists as at least tentatively "settling" some technical aspect of public issues, only to find the next day that other experts supply testimony for contrary conclusions. What results is confusion at best and cynicism about all authority at worst. Should witnessing audiences understand that science is *unfolding argumentation*—whatever else it may be—they would not be frustrated by controversial scientific claims. One function of this book is to show how the arguments that comprise scientific communication are designed and on what grounds they are and ought to be weighed as scientific claims.

Understanding the criteria governing the conventional rhetorical logic of science can help audiences judge whether specific applications are or are not situationally reasonable. Some commentators on science have expressed legitimate concern that the mystifying powers of scientific rhetoric can so confuse and intimidate the lay citizenry that they effectively become disenfranchised when trying to decide public issues.[4] Citizens would be better prepared to discern when scientific discussion leaves off and political exhortation begins if they comprehended what purposes are scientifically reasonable, what issues and problems scientists appropriately address, and what themes and values are conventionally used to legitimize claims as scientific. Indeed, citizens must also have practical understandings of historically evolved criteria for constraining reasonable democratic political rhetoric.[5] Only then could the comparative situational relevance of scientific and political argumentation be assessed. Without these understandings citizens are unable to discern when (1) a rhetoric of scientism is being used to secure extrascientific objectives or address issues not within the legitimate purview of reasonable scientific claim-making, and (2) extrascientific criteria of rhetorical reasonableness are used to legitimize claims as science. Only through learning the prominent rhetorical logics operating in public forums can citizens hope to remain fully enfranchised when exercising decision-making powers.

Finally, I hope to have shown the way for study of rhetorical invention carried on in other discursive communities. Any discursive community, as a community, has evolved some criteria for assessing the reasonableness of responses to recurring problems of communal concern. General inventional theory offers a field-invariant perspective for conducting field-by-field exploration of rhetorical logics customary within discursive communities. General inventional theory allows us to make explicit the organization of logical requirements and expectations governing the rhetoric of those fields. We can better understand a community's recurring exigences, norms for legitimate purposing, standard points for decision, and special *topoi* of thematic argument. Moreover, general inventional theory allows understanding of a discursive community's special rhetorical logic because the constituents of that logic emerge directly from the substantive concerns and expectations constraining the actual discursive practices of that community. We could not comprehend a discursive community's special rhetorical logic otherwise. It is because problems and values vary from one discursive community to another that the field-invariant perspective which general inventional theory offers is needed to extrapolate the special rhetorical logics governing the inventional practices of specific discursive communities.

The systematic approach to comprehending the rhetoric of science offered in this book can be extended heuristically to studies of rhetorical invention carried on in other self-identifying intellectual communities—philosophical, theological, literary/critical, and the like. In such communities, too, rhetorical purposing occurs under special, communally evolved constraints. The kinds of crucial *stases* differ, and, above all, standard *topoi* vary from one intellectual community to another. The general theory of rhetorical invention provides a field-invariant perspective for making specific analyses of such field-variant inventional practices.[6]

My objective has been to extend inventional analysis to science. Even here, however, variations on the basic inventional processes occur within specific sciences. I have given these no attention. Field-specific inventional processes are open to explorations that could show us the rhetorical processes peculiar to, for example, physics compared to biology or astronomy compared to chemistry. The special theory of invention for rhetoric of science provides a starting point for working out these similarities and differences.

Understanding the rhetoric of any intellectual community, within or outside the sciences and regardless of specific areas of specialty, requires that a central inventional question be raised and answered: How do members of self-identifying intellectual communities think differently and alike when creating reasonable discourse aimed at inducing adherence from their peers? My hope is that this book contributes to development of answers which expose grounds in common while recognizing pragmatic and intellectual differences. Only then could we hope to appreciate genuine diversity, avert submitting to artificial divisions, and seek to reintegrate currently fragmented disciplines without resorting to one or another form of intellectual imperialism.

NOTES

CHAPTER 1

1. Ernan McMullin, "Two Faces of Science," *The Review of Metaphysics* 27 (1974): 668.
2. Ibid., 668–70.
3. Karl R. Popper, *The Logic of Scientific Discovery* (New York: Harper & Row, 1968). My discussion of this and other critiques of positivism draws from McMullin's useful overview on 659–67.
4. Michael Polanyi, *Personal Knowledge: Towards a Post-Critical Philosophy* (Chicago: University of Chicago Press, 1958); and Norward Russell Hanson, *Patterns of Discovery* (Cambridge: Cambridge University Press, 1958).
5. Thomas S. Kuhn, *The Structure of Scientific Revolutions,* 2nd ed. (Chicago: University of Chicago Press, 1970).
6. Stephen Toulmin, *Human Understanding* (Princeton: Princeton University Press, 1972).
7. Paul Feyerabend, "Explanation, Reduction and Empiricism," in Herbert Feigl and Grover Maxwell, eds., *Minnesota Studies in the Philosophy of Science,* (Minneapolis: University of Minnesota Press, 1962), 3: 29. Also see Feyerabend, *Against Method: Outline of an Anarchistic Theory of Knowledge* (1975; rpt. London: Verso, 1978).
8. McMullin, "Two Faces of Science," 670–71.
9. Joseph Agassi called this historiographical approach "inductivist." Useful examples can be found in his "Towards an Historiography of Science," *History and Theory* 2 (1963): 1–117.
10. Central issues in traditional sociology of science are explained in Joseph Ben-David and Teresa A. Sullivan, "Sociology of Science," *Annual Review of Sociology* 1 (1975): 203–22. Also see Robert K. Merton's influential work, *The Sociology of Science: Theoretical and Empirical Investigations,* Norman W. Storer, ed. (Chicago: University of Chicago Press, 1973).
11. For instance, see Kuhn's influential *Structure of Scientific Revolutions.* Earlier but nonetheless exemplary works are Edwin Arthur Burtt, *The Metaphysical Foundations of Modern Physical Science* (New York: Harcourt, 1927); Alexandre Koyré, *From the Closed World to the Infinite Universe* (Baltimore: Johns Hopkins, 1957), and "Galileo and Plato," *Journal of the History of Ideas* 4 (1943): 400–28; and Edward J. Dijksterhuis, *The Mechanization of the World Picture* (Oxford: Clarendon Press, 1961).
12. Issues and research strategies are surveyed in H. M. Collins, "The Sociology of Scientific Knowledge: Studies of Contemporary Science," *Annual Review of Sociology* 9 (1983): 265–85. For an excellent example see Collins's "Son of Seven Sexes: The Social Destruction of a Physical Phenomenon," *Social Studies of Science* 11 (1981): 33–62.

13. Larry Laudan, "Two Puzzles About Science: Reflections About Some Crises in the Philosophy and Sociology of Science," *Minerva* 20 (1982): 253–54.
14. For evidence and examples see the essays in John S. Nelson, Allan Megill, and Donald N. McCloskey, eds., *The Rhetoric of the Human Sciences: Language and Argument in Scholarship and Public Affairs* (Madison: University of Wisconsin Press, 1987).
15. Stephen Toulmin, *The Uses of Argument* (Cambridge: Cambridge University Press, 1964), 218.
16. Maurice A. Finocchiaro, *Galileo and the Art of Reasoning: Rhetorical Foundations of Logic and Scientific Method* (Dordrecht: Reidel, 1980).
17. Floyd D. Anderson, rev. of *Galileo and the Art of Reasoning,* by Maurice A. Finocchiaro, *Philosophy and Rhetoric* 15 (1982): 136–38.
18. William James, *The Varieties of Religious Experience* (New York: Modern Library, 1929), Lecture X.
19. Walter B. Weimer, "Science as a Rhetorical Transaction: Toward a Nonjustificational Conception of Rhetoric," *Philosophy and Rhetoric* 10 (1977): 1–29. Also see his *Notes on the Methodology of Scientific Research* (Hillsdale, NJ: Erlbaum, 1979).

CHAPTER 2

1. The following works identify transformations in thought about rhetoric: Douglas Ehninger, "On Systems of Rhetoric," *Philosophy and Rhetoric* 1 (1968): 131–44; Chaim Perelman, "The New Rhetoric: A Theory of Practical Reasoning," in *The New Rhetoric and the Humanities: Essays on Rhetoric and Its Applications* (Dordrecht: Reidel, 1979), 1–42; Lawrence W. Rosenfield, "An Autopsy of the Rhetorical Tradition," in Lloyd F. Bitzer and Edwin Black, eds., *The Prospect of Rhetoric* (Englewood Cliffs, NJ: Prentice-Hall, 1971), 64–77; Karl R. Wallace, "The Fundamentals of Rhetoric," in Bitzer and Black, *The Prospect of Rhetoric,* 3–20.
2. P. Albert Duhamel, "The Function of Rhetoric as Effective Expression," *Journal of the History of Ideas* 10 (1949): 344.
3. For a modern example see Alfred Korzybski, *Science and Sanity: An Introduction to Non-Aristotelian Systems and General Semantics,* 4th ed. (1958; rpt. Lakeville, CT: International Non-Aristotelian Library, 1973). This concept of language can readily spill over into contempt for the arts of language, as illustrated in John Locke's criticism of rhetoric; see *An Essay Concerning Human Understanding,* Alexander Campbell Fraser, ed. (Oxford: Clarendon Press, 1894), vol. 2, bk. 3, ch. 10, sec. 34.
4. This view led Descartes to eschew the importance of studies in the language arts because he thought clarity of expression was a function of having ordered, coherent ideas in one's mind; the rules of rhetoric, in his view, have little to contribute to effectiveness of expression. See René Descartes, *Discourse on Method,* John Veitch, trans., in *The Rationalists* (Garden City, NY: Anchor Books, 1974), 43.
5. Wayne C. Booth, *Modern Dogma and the Rhetoric of Assent* (Chicago: University of Chicago Press, 1974), 135.
6. Michael C. Leff, "In Search of Ariadne's Thread: A Review of Recent

Literature on Rhetorical Theory," *Central States Speech Journal* 29 (1978): 84. Also see Charles P. Segal, "Gorgias and the Psychology of the Logos," *Harvard Studies in Classical Philology* 66 (1962): 99–155.

7. Kenneth Burke, *Rhetoric of Motives* (1950; rpt. Berkeley: University of California Press, 1969), 43.
8. Ibid., 43–44.
9. On this point Burke joins thinkers like Cassirer and Langer. See Burke's "Definition of Man," in his *Language as Symbolic Action: Essays on Life, Literature, and Method* (Berkeley: University of California Press, 1966), 3–24, esp. 23; Ernst Cassirer, *An Essay on Man* (New Haven: Yale University Press, 1944), esp. 23–26. For a profound discussion of the implications of this claim see Susanne K. Langer, *Philosophy in a New Key: A Study in the Symbolism of Reason, Rite and Art* (1942; rpt. New York: New American Library, 1951).
10. See Burke's reflections in "Definition of Man," 5.
11. Weinberg makes this point in connection with the aim of Zen Buddhism to achieve satori, or enlightenment; see Harry L. Weinberg, *Levels of Knowing and Existence: Studies in General Semantics,* 2nd ed. (1959; rpt. Lakeville, CT: Institute of General Semantics, 1973), 238.
12. This idea and its implications are developed more fully in Burke's "Terministic Screens," in *Language as Symbolic Action,* 44–62.
13. This image was originally developed by Erich Fromm to depict the emergence of humanity from purely instinctual, animal experience, once it developed the power of reason. This image especially applies to the transcendent power of the symbol. Once able to symbolize, humanity could *make* its world but could no longer return to some presymbolic world that simply is, as given; see Fromm, *The Art of Loving* (New York: Harper & Row, 1956), 6.
14. Burke, *Rhetoric of Motives,* 43.
15. Burke, "Terministic Screens," 45.
16. Chaim Perelman, *The Realm of Rhetoric,* William Kluback, trans. (Notre Dame, IN: University of Notre Dame Press, 1982), 61.
17. Richard M. Weaver, "Language Is Sermonic," in Robert E. Nebergall, ed., *Dimensions of Rhetorical Scholarship* (Norman: University of Oklahoma Department of Speech, 1963), 49–63.
18. See Burke's "Terministic Screens," 46, where he writes, "Much that we take as observations about 'reality' may be the spinning out of possibilities implicit in our particular choice of terms."
19. Burke explains that orientations give us a sense of relationships, or valuings, that constrain and shape interpretations of reality. See *Permanence and Change: An Anatomy of Purpose* (1954; rpt. Indianapolis, IN: Bobbs-Merrill, 1977), 35–36.
20. Ibid., 9–10.
21. In Sidney Hook, Paul Kurtz, and Miro Todorovich, eds., *The Philosophy of the Curriculum,* (Buffalo, NY: Prometheus Books, 1975), 96–97.
22. *Permanence and Change,* 84.
23. John B. Watson, *Behaviorism* (1924; rpt. New York: Norton, 1970), 6.
24. Burke, *Permanence and Change,* 85.

25. Ibid., 21, 203.
26. Robert K. Merton, "The Normative Structure of Science," in his *The Sociology of Science: Theoretical and Empirical Investigations,* Norman W. Storer, ed. (Chicago: University of Chicago Press, 1973), 270.
27. For illustration of how scientists can work from within different moral frameworks for assessing what is and what is not scientifically appropriate behavior see Mitroff's study of what he calls the "norms" and "counter-norms" for doing science; Ian I. Mitroff, *The Subjective Side of Science: A Philosophical Inquiry into the Psychology of the Apollo Moon Scientists* (Amsterdam: Elsevier, 1974).
28. Burke, *Permanence and Change,* 169.
29. Burke, *Rhetoric of Motives,* 55.
30. Ibid., 20–22.
31. *Rhetoric,* Rhys Roberts, trans. (New York: Modern Library, 1954), 1355b26.
32. Ibid., 1358a36–1359a6.
33. Lloyd F. Bitzer, "The Rhetorical Situation," *Philosophy and Rhetoric* 1 (1968): 6. For explanations of exigence, audience, and constraints see pp. 6–8.
34. Ibid., 3.
35. Donald C. Bryant, "Rhetoric: Its Function and Its Scope," *Quarterly Journal of Speech* 39 (1953): 413.
36. Burke, *Rhetoric of Motives,* 38–39. Also see Don M. Burks, "Persuasion, Self-Persuasion, and Rhetorical Discourse," *Philosophy and Rhetoric* 3 (1970): 109–19.
37. See Chaim Perelman and L. Olbrechts-Tyteca, *The New Rhetoric: A Treatise on Argumentation,* John Wilkinson and Purcell Weaver, trans. (Notre Dame, IN: University of Notre Dame Press, 1969), 5; and Perelman's *Realm of Rhetoric,* 4–5, 161–62.
38. Discussion of the representational roles of rhetors and audiences and of the communal standards that unite them is influenced in part by Lloyd F. Bitzer's "Rhetoric and Public Knowledge," in Don M. Burks, ed., *Rhetoric, Philosophy, and Literature: An Exploration* (West Lafayette, IN: Purdue University Press, 1978), 67–93.
39. Karl R. Wallace, "The Substance of Rhetoric: Good Reasons," *Quarterly Journal of Speech* 49 (1963): 239–40.
40. Ibid., 247–49. John S. Nelson has proposed that at least seven different kinds of "logics" or criteria of appropriateness and reasonableness operate in communicative inquiry; see his "Seven Rhetorics of Inquiry," in Nelson, et al., *The Rhetoric of the Human Sciences: Language and Argument in Scholarship and Public Affairs,* (Madison: University of Wisconsin Press, 1987), 407–34.
41. Chaim Perelman, "The Rational and the Reasonable," in *The New Rhetoric and the Humanities,* 117–18.
42. Ibid., 118.
43. For example, see John W. Ray, "Perelman's Universal Audience," *Quarterly Journal of Speech* 64 (1978): 361–75.
44. "The Rational and the Reasonable," 119–20. Also see Perelman's "The

New Rhetoric and the Rhetoricians: Remembrances and Comments," *Quarterly Journal of Speech* 70 (1984): 193–94.

45. Other useful treatments of "good reasons" may be found in Wallace's "Substance of Rhetoric," and Booth's *Modern Dogma and the Rhetoric of Assent*. Fisher correctly maintains that these treatments are less clear than the definition he offers; See Walter R. Fisher, "Toward a Logic of Good Reasons," *Quarterly Journal of Speech* 64 (1978): 376–84.
46. Fisher, "Toward a Logic of Good Reasons," 378.
47. Ibid., 378 n. 18.
48. Carroll Arnold, "What's Reasonable?" *Today's Speech* 19 (1971): 22.
49. *Ibid.*

CHAPTER 3

1. See as examples Douglas Ehninger, Bruce E. Gronbeck, Ray E. McKerrow, and Alan H. Monroe, *Principles and Types of Speech Communication,* 10th ed. (Glenview, IL: Scott, Foresman, 1986), 66–73; Otis M. Walter and Robert L. Scott, *Thinking and Speaking: A Guide to Intelligent Oral Communication,* 3rd ed. (New York: Macmillan, 1973), 24–29; and Eugene E. White, *Practical Public Speaking,* 4th ed. (New York: Macmillan, 1982), esp. 76–88.
2. See Edwin Black, *Rhetorical Criticism: A Study in Method* (1965; rpt. Madison: University of Wisconsin Press, 1978), esp. 22–27; Kenneth Burke, "The Rhetoric of Hitler's 'Battle,'" in his *Philosophy of Literary Form* (1941; rpt. Berkeley: University of California Press, 1973), 191–220; Richard B. Gregg, "The Ego Function of the Rhetoric of Protest," *Philosophy and Rhetoric* 4 (1971): 71–91; Dale G. Leathers, "Belief-Disbelief Systems: The Communicative Vacuum of the Radical Right," in G. P. Mohrmann, Charles J. Stewart, and Donovan J. Ochs, eds., *Explorations in Rhetorical Criticism* (University Park: Pennsylvania State University Press, 1973), 124–37; and Martin Maloney, "Clarence Darrow," in Marie Kathryn Hochmuth, ed., *A History and Criticism of American Public Address* (New York: McGraw-Hill, 1955), 3:262–312.
3. For instance, Richard B. Gregg extrapolated principles of cognitive processing that supposedly structure and constrain all symbolic inducement; see his *Symbolic Inducement and Knowing: A Study in the Foundations of Rhetoric* (Columbia: University of South Carolina Press, 1984).
4. George Campbell, *The Philosophy of Rhetoric,* Lloyd F. Bitzer, ed. (Carbondale: Southern Illinois University Press, 1963), xliii.
5. Ibid., 1–7, 71–94.
6. Ibid., 1.
7. Ibid., 95–96.
8. Alan H. Monroe and Douglas Ehninger, *Principles and Types of Speech,* 6th ed. (Glenview, IL: Scott, Foresman, 1967), 115–18. This classification is commonplace in public-speaking and persuasion textbooks. See Walter and Scott, *Thinking and Speaking,* 24–27, and White, *Practical Public Speaking,* 76–88. Also see Winston Lamont Brembeck and William Smiley

Howell, *Persuasion: A Means of Social Control* (Englewood Cliffs, NJ: Prentice-Hall, 1952), 296–99; and Robert T. Oliver, *Persuasive Speaking* (New York: Longmans, Green, 1950), 164–215.

9. For examples, see Monroe and Ehninger, *Principles and Types of Speech,* 418–20, 448–50, 469–70; Brembeck and Howell, *Persuasion,* 305–6; and Oliver, *Persuasive Speaking,* 164.
10. Aristotle, *Rhetoric,* Rhys Roberts, trans. (New York: Modern Library, 1954), 1355b26.
11. Ibid., 1358a36–1358b8.
12. Ibid., 1358b8–21.
13. Ibid., 1358b21–28.
14. Aristotle makes a similar point; ibid., 1358b29–1359a5.
15. Ibid., 1360b4–1365b20.
16. Ibid., 1366a23–1368a37.
17. Ibid., 1369b18–1372a3.
18. *Orator,* H. M. Hubbell, trans. (Cambridge, MA: Harvard University Press, 1962), 71. For a full discussion of propriety see 70–74. Also see *De oratore,* E. W. Sutton and N. Rackham trans. (Cambridge, MA: Harvard University Press, 1942), i.3.12.
19. See Richard McKeon, "Communication, Truth, and Society," *Ethics* 67 (1957): 89–99; Gerard A. Hauser and Donald P. Cushman, "McKeon's Philosophy of Communication: The Architectonic and Interdisciplinary Arts," *Philosophy and Rhetoric* 6 (1973): 213–18.
20. "Communication, Truth, and Society," 92–93.
21. For elaboration of McKeon's point see Hauser and Cushman, "McKeon's Philosophy," 217–18; and Donald P. Cushman and Phillip K. Tompkins, "A Theory of Rhetoric for Contemporary Society," *Philosophy and Rhetoric* 13 (1980): 47–52.
22. Sanborn made an excellent critique of psychologically based typologies of rhetorical ends, arguing that the term "persuasion" suffices to cover the entire range of efforts to influence thought and action through oral discourse. More recently Anderson argued, as did Sanborn, that any change in the beliefs, attitudes, and actions of persons is "persuasion." Sanborn further contended that viewing persuasion as a unitary process enhances attention to processes of invention and audience adaptation that would otherwise stagnate. George A. Sanborn, "The Unity of Persuasion," *Western Speech* 19 (1955): 175–83; and Kenneth E. Anderson, *Persuasion: Theory and Practice,* 2nd ed. (Boston: Allyn and Bacon, 1978).

CHAPTER 4

1. The following sources usefully introduce Greek and Roman ideas about *stasis:* George Kennedy, *Classical Rhetoric and Its Christian and Secular Tradition from Ancient to Modern Times* (Chapel Hill: University of North Carolina Press, 1980); *The Art of Rhetoric in the Roman World* (Princeton: Princeton University Press, 1972); Ray Nadeau, "Classical Systems of Stases in Greek: Hermagoras to Hermogenes," *Greek, Roman and Byzantine Studies* 2 (1959): 53–71; and "Hermogenes *On Stases:* A Translation with

an Introduction and Notes," *Speech Monographs* 31 (1964): 361–424; see esp. 369–86.

2. Otto Alvin Loeb Dieter, "Stasis," *Speech Monographs* 17 (1950): 345–69; see esp. 347–48.
3. Nadeau states that Dieter's attempt to trace the physical origins of *stasis* is insightful, but only as a clarifying analogy in relation to the four *stases* of Hermagoras and later rhetoricians; see Ray Nadeau, "Some Aristotelian and Stoic Influences on the Theory of Stases," *Speech Monographs* 26 (1959): 248.
4. Dieter, "Stasis," 350–51.
5. Ibid., 360–62.
6. Cicero's views on the nature of knowledge are discussed in Michael J. Buckley, S.J., "Philosophic Method in Cicero," *Journal of the History of Philosophy* 8 (1970): 143–54; Prentice A. Meador, Jr., "Rhetoric and Humanism in Cicero," *Philosophy and Rhetoric* 3 (1970): 1–12, and "Skeptic Theory of Perception: A Philosophical Antecedent of Ciceronian Probability," *Quarterly Journal of Speech* 54 (1968): 340–51; and Douglass F. Threet, "Rhetorical Function of Ciceronian Probability," *Southern Speech Communication Journal* 39 (1974): 309–21.
7. *De inventione,* H. M. Hubbell, trans. (Cambridge, MA: Harvard University Press, 1949), i.8.10.
8. H. M. Hubbell, introduction, *De inventione, De optimo genere oratorum, Topica* (Cambridge, MA: Harvard University Press, 1949), xii–xiii.
9. *De inventione,* i.8.11.
10. Ibid.
11. Ibid., i.9.12.
12. Ibid., i.11.16.
13. *Institutio oratoria,* H. E. Butler, trans. (Cambridge, MA: Harvard University Press, 1920), iii.6.7–9 (emphasis added). That there is always just *one* crucial point at issue is debatable at best.
14. Cicero, *De partitione oratoriae,* H. Rackham, trans. (Cambridge, MA: Harvard University Press, 1942), 29.101–30.104; *De inventione,* i.13.18–14.9.
15. *De inventione,* ii.4.14–ii.39.115.
16. *Topica,* H. M. Hubbell, trans. (Cambridge, MA: Harvard University Press, 1949), 21.81–82; *De partitione oratoriae,* 18.62–63.
17. This discussion is based on Buckley, "Philosophic Method in Cicero," 150–154; *De partitione oratoriae,* 18.61–67; *Topica,* 21.79–23.90
18. Cicero, *De oratore,* E. W. Sutton and H. Rackman, trans. (Cambridge, MA: Harvard University Press, 1942), i.31.138–41, ii.24.104; *Topica,* 25.93–94; *De inventione,* ii.4.12, ii.51.155; Quintilian, *Institutio,* iii.6.1, iii.6.81, iii.8.4.
19. *Institutio,* iii.6.80–82. In forensic rhetoric "legal" questions are concerned with disputes about written law and "rational" questions involve disputes about the alleged nature of facts or things (ibid., iii.5.4). Quintilian believes both are subsumed under the fundamental *stases* of conjecture, definition, and quality (ibid., iii.6.66–67, iii.6.86–89). Definite and indefinite questions are also encompassed by the three *stases.* Definite questions that turn

on specific circumstances of a person, time, or place (e.g., Should Cato marry?) are distinguished from indefinite philosophical questions that do not turn on specific circumstances (e.g., Should a man marry?). Definite questions logically presuppose indefinite questions, "since the *genus* is logically prior to the *species*" (ibid., iii.5.9); but systematic analysis of both kinds of question requires application of the three fundamental *stases* (ibid., iii.5.16). Full discussion of definite and indefinite questions is in ibid., iii.5.5–16.

20. See Lee S. Hultzén, "Status in Deliberative Analysis," in Donald C. Bryant, ed., *The Rhetorical Idiom: Essays in Rhetoric, Oratory, Language, and Drama* (New York: Russell and Russell, 1966), 97–123.
21. Hultzén's complete system includes twelve possible points at issue. Within dispute about any of the four deliberative considerations traditional types of clash might arise. For example, in argument about an *ill* there may be clash over whether the ill *exists,* how the ill should be *defined,* and whether there are extenuating circumstances that *qualify* the ill. These traditional *stases* may arise in dispute about the other three deliberative considerations as well. See ibid., 108–113.
22. Chaim Perelman and L. Olbrechts-Tyteca, *The New Rhetoric: A Treatise on Argumentation* (Notre Dame, IN: University of Notre Dame Press, 1969), 210–214.
23. Gerald M. Phillips, Douglas J. Pedersen, and Julia T. Wood, *Group Discussion: A Practical Guide to Participation and Leadership* (Boston: Houghton Mifflin, 1979), 127.
24. For an exploratory paper dealing with the use of *stasis* doctrine in group deliberation see Lawrence J. Prelli and Roger Pace, "*Stasis,* Good Reasons, and the Small Group," in Frans H. van Eemeren, Rob Grootendorst, J. Anthony Blair, and Charles A. Willard, eds., *Argumentation: Perspectives and Approaches* (Dordrecht: Foris, 1987), 255–65.
25. Since Scott and Lyman developed a theory of accounts, interpersonal communication scholars have generated an extensive literature on the subject; see Marvin B. Scott and Stanford M. Lyman, "Accounts," *American Sociological Review* 33 (1968): 46–62.
26. See Gerard A. Hauser, *Introduction to Rhetorical Theory* (New York: Harper & Row, 1986), 85–89; and John Adams, "An Explication and Presentation of an Expanded Typology of Forensic *Stases* for Use in the Study and Practice of Interpersonal Conflict," in Charles W. Kneupper, ed., *Oldspeak/Newspeak: Rhetorical Transformations* (Arlington, TX: Rhetoric Society of America, 1985), 227–40.
27. Hauser, *Introduction,* 87.
28. Ibid.
29. This is not to say that each of these lines of thought is only applicable to qualitative *stases.* For instance, "denial of injury" might yield useful themes for classifying behaviors in ways that engage definitional *stases.* Generally, however, people make justifications and excuses when (1) they admit to performing the behavior, and (2) the locus of conflict concerns the *quality* of the behavior so performed.
30. Adams applies legal *stasis* categories to interpersonal conflicts, suggesting

that we are all potentially lawyers when we enter into interpersonal relationships: we often find ourselves *accusing* our intimate associates with having engaged in improper conduct toward us or *defending* ourselves from similar charges of impropriety. My present point is that *stasis* analysis can be extended usefully to interpersonal confusions as well as conflicts.

31. Richard McKeon, "Communication, Truth, and Society," *Ethics* 67 (1957): 89–99; and Gerard A. Hauser and Donald P. Cushman, "McKeon's Philosophy of Communication: The Architectonic and Interdisciplinary Arts," *Philosophy and Rhetoric* 6 (1973): 213–18.
32. Methodological features discussed here are indicated in McKeon, "The Uses of Rhetoric in a Technological Age: Architectonic Productive Arts," in Lloyd F. Bitzer and Edwin Black, eds., *The Prospect of Rhetoric* (Englewood Cliffs, NJ: Prentice-Hall, 1971), 47–48, 54–58; and "The Methods of Rhetoric and Philosophy: Invention and Judgment," in Luitpold Wallach, ed., *The Classical Tradition* (Ithaca, NY: Cornell University Press, 1966), 368–71.
33. Cushman and Tompkins sought to apply McKeon's perspective in *stasis* analysis of practical rhetorical situations where parties with varying ideological perspectives differ about the nature of and interrelationships among the elements at work in any given rhetorical situation; see "A Theory of Rhetoric for Contemporary Society," *Philosophy and Rhetoric* 13 (1980): 57–58.
34. *De inventione,* i.8.10.
35. Nadeau, "Classical Systems of Stases in Greek," 61.
36. Ibid., 68.
37. *De inventione,* i.10.14.

CHAPTER 5

1. Hugh Blair, *Lectures on Rhetoric and Belles Lettres* (1783; rpt. Carbondale: Southern Illinois University Press, 1965), 2: 181–82.
2. In the early eighteenth century Vico lamented the trend in education that emphasized logical analysis or "philosophical criticism" at the expense of the "art of topics." In his view, "This is harmful, since the invention of arguments is by nature prior to the judgment of their validity, so that in teaching, invention should be given priority over philosophical criticism"; Giambattista Vico, *On the Study Methods of Our Time,* Elio Gianturco, trans. (Indianapolis: Bobbs-Merrill, 1965), 14.
3. Toulmin rejects formal logic as a model for practical argument largely because it does not apply to the ways people argue about *substantive* matters; see Stephen Toulmin, *The Uses of Argument* (Cambridge: Cambridge University Press, 1964), esp. chs. 2 and 3.
4. Polanyi contends that all attempts to formulate the process of inductive inference fail because they neglect the centrality of a "framework of personal judgment" to all selection and verification of hypotheses. What constitutes the facts or specific instances from which inductive inferences are drawn depends on conceptions or values formulated tacitly or explicitly by those

making the inferences; see Michael Polanyi, *Personal Knowledge* (Chicago: University of Chicago Press, 1958), esp. 29–30.

5. Historians of logic have overlooked, minimized, or neglected informal topical method in favor of examining the historical development of formal syllogistic structures. Studies of Aristotelian logic illustrate this. Kneal and Kneale imply that Aristotle's informal logic of the *Topics*—a method for dialectical argument with an interlocutor and not a structure for ordering scientific argument—is an inferior precursor to the pristine formal logic of the *Prior Analytics*. Ross claims explicitly that the *Analytics* made the *Topics* an out-of-date treatise. Bochenski, although recognizing the importance of research into the topics, gives only cursory treatment. See William Kneale and Martha Kneale, *The Development of Logic* (Oxford: Oxford University Press, 1962), 33; W. D. Ross, *Aristotle* (London: Methuen, 1923), 59; I. M. Bochenski, *A History of Formal Logic* (Notre Dame, IN: University of Notre Dame Press, 1961), 28, 49–50
6. Useful historical treatments of topical theory are available. Three books by George Kennedy are of general use: *The Art of Persuasion in Greece* (Princeton: Princeton University Press, 1963); *The Art of Rhetoric in the Roman World* (Princeton University Press, 1972); and *Classical Rhetoric and Its Christian and Secular Tradition from Ancient to Modern Times* (Chapel Hill: University of North Carolina Press, 1980). More specialized but explanatory works include the following: Otto Bird, "The Tradition of the Logical Topics: Aristotle to Ockham," *Journal of the History of Ideas* 23 (1962): 307–23; Ernesto Grassi, *Rhetoric as Philosophy: The Humanist Tradition* (University Park: Pennsylvania State University Press, 1980); Richard McKeon, "Creativity and the Commonplace," *Philosophy and Rhetoric* 6 (1973): 199–210; and Karl R. Wallace, *Francis Bacon on Communication and Rhetoric* (University of North Carolina Press, 1943), esp. ch. 3.
7. Aristotle, *Rhetoric,* Rhys Roberts, trans. (New York: Modern Library, 1954), 1356b26–34.
8. An excellent illustration of a topical system is Mortimer Adler's *Syntopicon*. The *Syntopicon* is a collection of topics that helps us to find information about important issues and positions taken about them in a collection of 443 works contained in the 54-volume set of the Great Books of the Western World. Adler composed a list of 102 basic ideas that was later subdivided by a staff of indexers into 3,000 topics. Information about a subject can be discovered by considering a list of related topics each of which contains references to pertinent passages in the set of books. Cross-references indicate similar or related topics that, once considered, can further enrich understanding of the subject. See Mortimer J. Adler and William Gorman, eds., *The Great Ideas: A Syntopicon of Great Books of the Western World,* 2 vols. (Chicago: Encyclopaedia Britannica, 1952). Also see the useful exposition of this work in Edward P. J. Corbett, *Classical Rhetoric for the Modern Student,* 2nd ed. (New York: Oxford University Press, 1971), 198–200.
9. Richard Young, "Invention: A Topographical Survey," in *Teaching Com-*

position: 10 Bibliographical Essays, Gary Tate, ed. (Fort Worth: Texas Christian University Press, 1976), 2.

10. *Rhetoric,* 1366a34–1366b22.
11. Ibid., 1355b10.
12. Ibid., 1396a4–1396b11.
13. William F. Nelson, "*Topoi:* Functional in Human Recall," *Speech Monographs* 37 (1970): 121–26; Dominic A. Infante, "The Influence of a Topical System on the Discovery of Arguments," *Speech Monographs* 38 (1971): 125–28; John L. Petelle and Richard Maybee, "Items of Information Retrieved as a Function of Cue System and Topical Area," *Central States Speech Journal* 25 (1974): 190–97; William Nelson, John L. Petelle, and Craig Monroe, "A Revised Strategy for Idea Generation in Small Group Decision Making," *Speech Teacher* 23 (1974): 191–96.
14. Students of Aristotle have difficulty characterizing the various kinds of *topoi* he introduced. For instance, Grimaldi claims special *topoi* are concerned with the "matter" and general *topoi* are concerned with the "forms" of argument. Conley argues that the two kinds of *topoi* should be distinguished according to relative degrees of "field dependence/field invariance." Regardless of what Aristotle really meant, I take the position that "special" shall designate field-dependent, substantive topics and "general" shall designate topics possibly useful in any field and on any subject. Thomas M. Conley, "'Logical Hylomorphism' and Aristotle's *Koinoi Topoi,*" *Central States Speech Journal* 29 (1978): 92–97; and William Grimaldi, S.J., "The Aristotelian Topics," *Traditio* 14 (1958): 1–16.
15. Grimaldi, "Aristotelian Topics," 10.
16. Aristotle identifies this feature of special *topoi* in *Rhetoric,* 1358a17–20.
17. Ibid., 1358a2–9; 1358a23–26. Also see Grimaldi, "Aristotelian Topics," 10–11.
18. Literature illustrating topical ways of thinking should include the work of the sociologist Nisbet and the relatively older works of the historians Becker and Lovejoy, even though these scholars do not explicitly develop and apply a concept of topics. Nisbet's analysis of sociology as an art form is a striking illustration of topical thinking applied to an academic field of study. Lovejoy's method of analyzing the history of a philosophical doctrine for its constituent "unit-ideas" is similar to exploration of fundamental topics whose varied applications and arrangements characterize that body of thought's development. Becker's description of the "unobtrusive words" that represent the "climate of opinion" for an age has strong resemblance to central qualities of rhetorical *topoi.* See Carl L. Becker, *The Heavenly City of the Eighteenth-Century Philosophers* (New Haven: Yale University Press, 1932); Arthur O. Lovejoy, *The Great Chain of Being* (Cambridge, MA: Harvard University Press, 1936); Robert Nisbet, *Sociology as an Art Form* (London: Oxford University Press, 1976).
19. Ruth Anne Clark and Jesse G. Delia, "*Topoi* and Rhetorical Competence," *Quarterly Journal of Speech* 65 (1979): 187–206; Charles W. Kneupper and Floyd D. Anderson, "Uniting Wisdom and Eloquence: The Need for Rhetorical Invention," *Quarterly Journal of Speech* 66 (1980): 313–26;

Karl R. Wallace, "*Topoi* and the Problem of Invention," *Quarterly Journal of Speech* 58 (1972): 387–95; Wilhelm Hennis, "Political Science and the Topics," *Graduate Faculty Philosophy Journal* 7 (1978): 35–77.

20. For examples of efforts working in this direction see Nancy S. Struever, "Topics in History," *History and Theory* 19 (1980): 66–79; Donald N. McCloskey, "The Literary Character of Economics," *Daedalus* 113 (1984): 97–119; "The Rhetoric of Economics," *Journal of Economic Literature* 21 (1983): 481–517; and *The Rhetoric of Economics* (Madison: University of Wisconsin Press, 1985). Some of the essays contained in Nelson et al. engage in topical ways of thinking about academic fields of inquiry, but do so less self-conciously; see John S. Nelson, Allan Megill, and Donald N. McCloskey, eds. *The Rhetoric of the Human Sciences: Language and Argument in Scholarship and Public Affairs* (Madison: University of Wisconsin Press, 1987).
21. Nisbet, *Sociology as an Art Form,* 31.
22. Ibid., 37.
23. Ibid., 37–39.
24. Ibid., 39. The special themes of sociology are explored in detail in Nisbet's, *The Sociological Tradition* (New York: Basic Books, 1966).
25. McCloskey extrapolates these and other *topoi* in *Rhetoric of Economics;* see esp. 57–60, 69–74, 102–112. McCloskey calls classical figures of speech drawn from Lanham's *Handlist of Rhetorical Terms* "common topics" (pp. 126–30). These he contrasts with special topics peculiar to economics rhetoric (pp. 130–33). Also see Richard A. Lanham, *A Handlist of Rhetorical Terms: A Guide for Students of English Literature* (Berkeley: University of California Press, 1968), sec. 3.11 and sections on figures of speech.
26. Craig Kallendorf and Carol Kallendorf, "A New Topical System for Corporate Speechwriting," *Journal of Business Communication* 21 (1984): 3–14.
27. See Donald C. Bryant and Karl R. Wallace, *Fundamentals of Public Speaking,* 4th ed. (New York: Meredith, 1947); Corbett, *Classical Rhetoric for the Modern Student;* and John F. Wilson and Carroll C. Arnold, *Public Speaking as a Liberal Art,* 5th ed. (Boston: Allyn and Bacon, 1983).
28. Edward D. Steele and W. Charles Redding, "The American Value System: Premises for Persuasion," *Western Speech* 26 (1962): 83–91.
29. Ibid., 84.
30. For an analysis of Nixon's expense-fund speech using Steele and Redding's list of political *topoi* see Henry E. McGuckin, Jr., "A Value Analysis of Richard Nixon's 1952 Campaign-Fund Speech," *Southern Speech Communication Journal* 33 (1968): 259–69.
31. McKeon discusses the processes of creative transformation in "Creativity and the Commonplace," 199–210.
32. For provocative discussion of this point see Jacques Ellul, *A Critique of the New Commonplaces* (New York: Knopf, 1968), 3–27.

CHAPTER 6

1. Philosophers of science have been especially critical of Kuhn's work. Criticisms of Kuhn's paradigm theory cluster around three central issues: first,

whether the paradigm concept itself is ambiguous or vacuous, and therefore of little assistance to our understanding of scientific activity; second, whether the distinction between "normal" and "revolutionary" science has been adequately drawn; and third, whether Kuhn's contention that conflicting paradigms are "incommensurable" makes paradigm change an irrational process. See Frederick Suppe, ed., *The Structure of Scientific Theories,* 2nd ed. (Chicago: University of Illinois Press, 1977); and Imre Lakatos and Alan Musgrave, eds., *Criticism and the Growth of Knowledge* (London: Cambridge University Press, 1970). In the Suppe volume, session 6 of the symposium, which includes Kuhn's "Second Thoughts on Paradigms" and Suppe's "Exemplars, Theories and Disciplinary Matrixes," is especially useful as an overview of philosophical problems with Kuhn's work. The earlier Lakatos and Musgrave volume is devoted entirely to philosophical critique of Kuhn's claims.

2. For example, see Michael Heidelberger, "Some Intertheoretic Relations Between Ptolemean and Copernican Astronomy," in Gary Gutting, ed., *Paradigms and Revolutions: Appraisals and Applications of Thomas Kuhn's Philosophy of Science* (Notre Dame, IN: University of Notre Dame Press, 1980), 271–83.
3. Margaret Masterman, "The Nature of a Paradigm," in Lakatos and Musgrave, *Criticism,* 61–66. For a less sympathetic account of the many senses of "paradigm" see Dudley Shapere, "The Structure of Scientific Revolutions," *Philosophical Review* 73 (1964): 383–94.
4. *The Structure of Scientific Revolutions,* 2nd ed. (Chicago: University of Chicago Press, 1970), 175.
5. In early discussions of normal science Kuhn used the term "consensus" to designate this sociological dimension of the normal scientific enterprise; he used the term "paradigm" to designate the concrete solutions and standard examples used in scientific education. Later the paradigm concept displaced all talk of consensus and took on the more global proportions attributed to the sociological paradigm. In retrospect, Kuhn pointed to the "exemplary" sense as that which is more appropriately a paradigm. For a discussion of how the paradigm concept took on life of its own see Kuhn's preface to his *The Essential Tension: Selected Studies in Scientific Tradition and Change* (Chicago: University of Chicago Press, 1977), esp. xviii–xx.
6. *Structure of Scientific Revolutions,* 182; also see "Second Thoughts on Paradigms," *Essential Tension,* 297. All references to "Second Thoughts" will be to *Essential Tension.* It is also available in Suppe, *Structure of Scientific Theories.*
7. *Structure of Scientific Revolutions,* 182–84; also see "Second Thoughts," 298–301.
8. "Second Thoughts," 297–98; also see *Structure of Scientific Revolutions,* 184.
9. *Structure of Scientific Revolutions,* 184–86.
10. Ibid., viii.
11. Barry Barnes, *T. S. Kuhn and Social Science* (New York: Columbia University Press, 1982), 51–52.
12. This discussion is based on Kuhn, "Second Thoughts," 305–08; *Structure*

of Scientific Revolutions, 187–91.

13. Douglas Lee Eckberg and Lester Hill, Jr., "The Paradigm Concept and Sociology: A Critical Review," in Gutting, *Paradigms and Revolutions,* 120.
14. Ibid., 131.
15. Introduction, Gutting, *Paradigms and Revolutions,* 13.
16. Paul N. Campbell, "The *Personae* of Scientific Discourse," *Quarterly Journal of Speech* 61 (1975): 393.
17. Karin D. Knorr-Cetina, *The Manufacture of Knowledge: An Essay on the Constructivist and Contextual Nature of Science* (New York: Pergamon Press, 1981).
18. Ibid., 42.
19. Chaim Perelman and L. Olbrechts-Tyteca, *The New Rhetoric,* John Wilkinson and Purcell Weaver, trans. (Notre Dame, IN: University of Notre Dame Press, 1969), 5.
20. Henry W. Johnstone, Jr., *The Problem of the Self* (University Park: Pennsylvania State University Press, 1970), 121.
21. *Structure of Scientific Revolutions,* 94.
22. Ibid., 203.
23. Johnstone makes a philosophical argument in support of the view that the very process of critiquing any position other than one's own requires an understanding of that position on its own terms. See Henry W. Johnstone, Jr., *Validity and Rhetoric in Philosophical Argument: An Outlook in Transition* (University Park, PA: Dialogue Press of Man and World, 1978), esp. ch. 8.
24. "Second Thoughts on Paradigms," 294–98.
25. *Structure of Scientific Revolutions,* 178–79; "Second Thoughts," 295 n. 4; and "Reflections on My Critics," in Lakatos and Musgrave, *Criticism,* 272 n. 1.
26. Thomas S. Kuhn, "The Function of Dogma in Scientific Research," in A. C. Crombie, ed., *Scientific Change* (London: Heinemann, 1963), 349; also see *Structure of Scientific Revolutions,* 136–143.
27. Barnes, *T. S. Kuhn and Social Science,* 18–19.
28. *Structure of Scientific Revolutions,* 24.
29. Ibid., 35–42.
30. Kuhn, "The Essential Tension: Tradition and Innovation in Scientific Research," *Essential Tension,* 233. For further discussion see *Structure of Scientific Revolutions,* 23–34.
31. Practitioners of the disciplines of radiochemistry, nuclear physics, astrophysics, and neutrino physics attempted to solve an anomaly they all shared. When interviewed, members from each specialty tried to attribute the anomaly to an error made by practitioners in some other specialty, while defending their own specialty as more firmly grounded. This illustrates interparadigmatic discourse that makes use of such scientific values as accuracy, simplicity, and experimental competence in arguing for superiority of one paradigmatic orientation over others. See Trevor J. Pinch, "The Sun-Set: The Presentation of Certainty in Scientific Life," *Social Studies of Science* 11 (1981): 131–58.

32. Kuhn's theory of paradigmatic science seems ill equipped for accommodating this kind of interfield scientific discourse. See Lindley Darden and Nancy Mull, "Interfield Theories," *Philosophy of Science* 44 (1977): 43–64.
33. Ibid., 51–54. Darden and Mull suggest that three conditions must be present before interfield theories can be forged. Interfield theories are generated (1) when two fields share an interest in explaining different aspects of the same phenomenon; (2) when there exists background knowledge relating the two fields; (3) when questions arise in each field which are not answerable using the concepts and techniques of that field (ibid., 50).
34. *Structure of Scientific Revolutions,* 153–57.
35. "Essential Tension," 236.
36. *Structure of Scientific Revolutions,* 65.
37. Ibid., 52–53, 67–68, 82–83.
38. Quoted in ibid., 83–84.
39. Ibid., 84–85.
40. For full discussion of revolutionary paradigm change with reference to these and other examples see ibid., chs. 7–10.
41. For instance, Campbell attributes Darwin's rhetorical success to the skillful use of appeals that drew creatively upon a prevailing cultural grammar; John Angus Campbell, "Charles Darwin: Rhetorician of Science," in John S. Nelson, et al., *The Rhetoric of the Human Sciences: Language and Argument in Scholarship and Public Affairs* (Madison: University of Wisconsin Press, 1987), 69–86; and "Scientific Revolution and the Grammar of Culture: The Case of Darwin's *Origin,*" *Quarterly Journal of Speech* 72 (1986): 351–76.
42. See Stephen Jay Gould, *Ever Since Darwin: Reflections in Natural History* (New York: Norton, 1977), 41; and Campbell, "Scientific Revolution and the Grammar of Culture," 363. In Gould's view, Darwin's creative insight into how evolution works arose from analogical extension of Adam Smith's economics to biology. On this point see Gould's account in *The Panda's Thumb* (New York: Norton, 1980), 65–68. Also see the article on which Gould bases his reflections: Silvan S. Schweber, "The Origin of the *Origin* Revisited," *Journal of the History of Biology* 10 (1977): 229–316.
43. Kuhn, "The Historical Structure of Scientific Discovery," in *Essential Tension,* 165–77. Also see *Structure of Scientific Revolutions,* 52–65.
44. Augustine Brannigan, *The Social Basis of Scientific Discoveries* (Cambridge: Cambridge University Press, 1981), 77.
45. Ibid., 175.
46. Ibid., 77–78.
47. As a sample, see the following: Marlan Blissitt, *Politics in Science* (Boston: Little, Brown, 1972); Don K. Price, *The Scientific Estate* (Cambridge, MA: Harvard University Press, 1965); Edwin Mansfield, *The Economics of Technological Change* (New York: Norton, 1968); Jean-Jacques Salomon, *Science and Politics,* Noel Lindsay, trans. (London: Macmillan, 1973); Harvey M. Sapolsky, "Science, Technology and Military Policy," in Ina Spiegel-Rosing and Derek de Solla Price, eds., *Science, Technology and Society: A Cross-Disciplinary Perspective* (London: Sage, 1977), 443–71;

Leslie Sklair, *Organized Knowledge: A Sociological View of Science and Technology* (London: Hart-Davis, MacGibbon, 1973); and Alvin M. Weinberg, *Reflections on Big Science* (Cambridge, MA: MIT Press, 1967).

48. *Structure of Scientific Revolutions,* 168–69. Also see Zuckerman's excellent review of sociological literature on factors that constrain problem choice in science: Harriet Zuckerman, "Theory Choice and Problem Choice in Science," *Sociological Inquiry* 48 (1978): 73–86.
49. P. B. Medawar discusses this pragmatic standard in *The Art of the Soluble* (London: Methuen, 1967), esp. 7.
50. Bruno Latour and Steve Woolgar, *Laboratory Life: The Social Construction of Scientific Facts* (London: Sage, 1979), 176.
51. Ibid., 238.
52. Medawar argues that scientific articles actively misrepresent the processes of thought involved when scientists make discoveries. The false image is that scientific discoveries are based on a logical procedure of inductive inference from unprejudiced stable observations. Rather, discoveries involve "inspirational" acts of hypothesis formation, which shape expectations about what will be "discovered" through scientific inquiry; see P. B. Medawar, "Is the Scientific Paper Fraudulent?" *Saturday Review,* 1 Aug. 1964, 42–43.
53. The stylistic conventions mentioned here are discussed in S. Michael Halloran, "The Birth of Molecular Biology: An Essay in the Rhetorical Criticism of Scientific Discourse," *Rhetoric Review* 3 (1984): 74–75.
54. *Publication Manual of the American Psychological Association,* 2nd ed. (Washington, DC: American Psychological Association, 1974), 28.
55. Halloran and Bradford claim that instruction in the principles of classical rhetoric can help beginning engineers and scientists write more forcefully. Technical writers can become more conscious and critical of their specialty's conventional writing norms through studying stylistic figures. Exploitation of opportunities for creative individual expression is encouraged, but not to the point of neglecting writing as *adapted* discourse. With proper rhetorical training, beginning engineers and scientists "will recognize the writing norms of a given field as constraints to be taken seriously, but not slavishly. He or she will bring to his or her work the quality that Cicero called tact *(aptus),* an attunement to the complexities of the given situation allowing him or her to judge when to conform and when to be creative" (192); see S. Michael Halloran and Annette Norris Bradford, "Figures of Speech in the Rhetoric of Science and Technology," in Robert J. Connors, Lisa S. Ede, and Andrea A. Lunsford, eds., *Essays on Classical Rhetoric and Modern Discourse* (Carbondale: Southern Illinois University Press, 1984), 179–92.
56. *Manufacture of Knowledge,* 6.
57. Aristotle, *Rhetoric,* W. Rhys Roberts, trans. (New York: Modern Library, 1954), 1356a5.
58. A systematic rendering of Merton's ideas appears in "Science and Technology in a Democratic Order," *Journal of Legal and Political Sociology* 1 (1942): 115–26. I am using the reprint of this essay, entitled "The Normative Structure of Science," in Robert K. Merton, *The Sociology of Science: Theoretical and Empirical Investigations,* Norman W. Storer, ed.,

(Chicago: University of Chicago Press, 1973), 267–78. All references to Merton's articles will be to this collection of essays.

59. "Normative Structure of Science," 268–69. Merton's scientific ethos is the sociological complement to the epistemological position that Scheffler called the "standard view." The constitutive norms of the scientific ethos are alleged to ensure that scientists engage in behaviors that help them to fulfill their functional roles in achieving the institutional goal of science: extension of certified knowledge. See Israel Scheffler, *Science and Subjectivity* (Indianapolis: Bobbs-Merrill, 1967). For a critique of the philosophical position upon which Merton's view of scientific ethos rests, see Michael Mulkay, *Science and the Sociology of Knowledge* (London: Allen and Unwin, 1979).
60. "Normative Structure of Science," 270.
61. Ibid., 273–75. Following Barber, I am here substituting the term "communality" for what Merton called "intellectual communism" in order to avoid obscuring the meaning of this norm with political and ideological associations that might be invoked by the latter expression. See Bernard Barber, *Science and the Social Order* (1952; rpt. Westport, CT: Greenwood, 1978), 268 n. 7.
62. "Normative Structure of Science," 275–77. I am relying on Barber's concise definition; *Science and the Social Order*, 92.
63. "Normative Structure of Science," 277–78.
64. "Priorities in Scientific Discovery," *Sociology of Science*, esp. 293–305.
65. For examples of research strongly supportive of Merton's position see Jonathan R. Cole and Stephen Cole, *Social Stratification in Science* (Chicago: University of Chicago Press, 1973); Jonathan R. Cole, *Fair Science: Women in the Scientific Community* (New York: Free Press, 1979); Jerry Gaston, *The Reward System in British and American Science* (New York: Wiley, 1978); Harriet Zuckerman and Robert K. Merton, "Institutionalized Patterns of Evaluation in Science," in Merton, *Sociology of Science*, 460–96; and Harriet Zuckerman, "Stratification in American Science," *Sociological Inquiry* 40 (1970): 235–57. Gaston contends that Blissitt offers positive evidence for Merton's norms, but Mulkay seems to think otherwise. See Blissitt, *Politics in Science*, 72–73; Gaston, *Reward System*, 182–84; and Mulkay, *Science and the Sociology of Knowledge*, 68.
66. Ian I. Mitroff, *The Subjective Side of Science* (Amsterdam: Elsevier, 1974).
67. For lists containing these and other norms and counternorms see Mitroff, *Subjective Side of Science*, 79; and Ian I. Mitroff and Richard O. Mason, *Creating a Dialectical Social Science: Concepts, Methods, and Models* (Dordrecht: Reidel, 1981), 147–48.
68. See esp. S. B. Barnes and R. G. A. Dolby, "The Scientific Ethos: A Deviant Viewpoint," *Archives Européennes De Sociologie* 11 (1970): 3–25; and Mulkay, *Science and the Sociology of Knowledge*, 63–73.
69. For an overview of central issues in this dispute see Nico Stehr, "The Ethos of Science Revisited: Social and Cognitive Norms," *Sociological Inquiry* 48 (1978): 172–96.
70. *Rhetoric*, 1356a5–13.

71. In 1958 Polanyi made the point that eminent scientists with established records of having performed successful scientific work were more likely to gain endorsement for problematic technical claims. According to Zuckerman, she and Merton found evidence of this when they examined referees' reasons for recommending or rejecting submissions to the prestigious physics journal, *Physical Review*. Referees were more likely to endorse unorthodox ideas when authored by established and known scientists. See Michael Polanyi, *Personal Knowledge,* (Chicago: University of Chicago Press, 1958); Merton and Zuckerman, "Institutionalized Patterns of Evaluation," esp. 476–91; and Zuckerman, "Theory Choice and Problem Choice," 70.
72. Mulkay, *Science and the Sociology of Knowledge,* 71.
73. Rhetorical analysis cannot proceed fruitfully without recognizing that scientific discourse is addressed to an audience. As Anderson pointed out in his review of Finocchiaro's thought-provoking study of Galileo's *Dialogue Concerning the Two Chief World Systems,* Finocchiaro's work is blemished by its failure to recognize explicitly that the *Dialogue* was created for an audience at a particular time in the history of ideas; see Floyd D. Anderson, rev. of *Galileo and the Art of Reasoning: Rhetorical Foundations of Logic and Scientific Method,* by Maurice A. Finocchiaro, *Philosophy and Rhetoric* 15 (1982): 137.
74. This story is told by Henry Harris, "Rationality in Science," in A. F. Heath, ed., *Scientific Explanation: Papers Based on Herbert Spencer Lectures Given in the University of Oxford* (Oxford: Clarendon Press, 1981), 39–40.
75. For elaboration of this idea see Diana Crane, *Invisible Colleges: Diffusion of Knowledge in Scientific Communities* (Chicago: University of Chicago Press, 1972).
76. *Manufacture of Knowledge,* 7.
77. *Structure of Scientific Revolutions,* 168.
78. Dolby describes a hierarchy of audiences to which scientists address special kinds of arguments. They include, in descending order of importance, a scientist's own self, like-minded fellow specialists, specialists who do not share a scientist's approach to the specialty area, students, scientists in other specialties, nonscientific groups (industrial employers and political patrons), and the general public; see R. G. A. Dolby, "Sociology of Knowledge in Natural Science," *Science Studies* 1 (1971): 18–21.
79. *Structure of Scientific Revolutions,* 168.
80. One of Kuhn's staunchest opponents on the role of logic in scientific activity is Karl Popper. See his "Normal Science and Its Dangers," in Lakatos and Musgrave, *Criticism,* esp. 55–58. Also see Kuhn's "Reflections on My Critics," in ibid., esp. 260–62.
81. Paul Feyerabend, *Against Method* (1975; rpt. London: Verso, 1978).
82. Kuhn has tried to address this problem in *Structure of Scientific Revolutions,* 198–204; and "Reflections on My Critics," 259–70, 275–77.
83. Kuhn makes this functional distinction in "Objectivity, Value Judgment, and Theory Choice," *Essential Tension,* 331; and "Reflections on My Critics," 262.

84. Kuhn, "Reflections on My Critics," 260–261; *Structure of Scientific Revolutions,* 199.
85. See H. Putnam, "Philosophers and Human Understanding," in Heath, *Scientific Explanation,* 105–11.
86. Ibid., 107.
87. Ibid., 110–11.
88. Maurice A. Finocchiaro, *Galileo and the Art of Reasoning: Rhetorical Foundations of Logic and Scientific Method* (Dordrecht: Reidel, 1980), 46, 65.
89. Ibid., 435.
90. Ibid., 434. Analysis of this rhetorical dimension of the *Dialogue* is on pp. 6–22.
91. Ibid., 46.
92. Ibid., 434.
93. Ibid., 3–4.
94. Ibid., 46.
95. Ibid., 46–65.
96. Ibid., 65.
97. Ibid., 290–91.
98. Ibid., 24, 44–45, 188–200. Also see his "Logic and Rhetoric in Lavoisier's Sealed Note: Toward a Rhetoric of Science," *Philosophy and Rhetoric* 10 (1977): 111–22.
99. *Structure of Scientific Revolutions,* 94.
100. "Logic of Discovery or Psychology of Research?" in Lakatos and Musgrave, *Criticism,* 22.
101. Walter B. Weimer, "Science as a Rhetorical Transaction," *Philosophy and Rhetoric* 10 (1977): 13.
102. Ibid., 20.
103. Herbert W. Simons, "Are Scientists Rhetors in Disguise? An Analysis of Discursive Processes Within Scientific Communities," in Eugene E. White, ed., *Rhetoric in Transition: Studies in the Nature and Uses of Rhetoric* (University Park: Pennsylvania State University, 1980), 127.
104. Michael A. Overington, "The Scientific Community as Audience: Toward a Rhetorical Analysis of Science," *Philosophy and Rhetoric* 10 (1977): 144.
105. Overington does identify what I have called "general topics" at work in sociological discourse, but as field-independent topics they do not reveal the special themes that are used to legitimate claims *as scientific* in sociology; see ibid., 157–59.
106. As an example see John S. Nelson, "Seven Rhetorics of Inquiry: A Provocation," in Nelson et al., *Rhetoric of the Human Sciences,* 407–34; and Michael Calvin McGee and John R. Lyne, "What Are Nice Folks Like You Doing in a Place Like This? Some Entailments of Treating Knowledge Claims Rhetorically," in ibid., 398–400.
107. Handbooks on scientific writing and speaking neglect inventional considerations involved in making and adapting discourse to situated audiences. For an example of standard handbooks on making scientific discourse, see Vernon Booth, *Communicating in Science: Writing and Speaking* (Cambridge: Cambridge University Press, 1984).

CHAPTER 7

1. Thomas S. Kuhn, *The Structure of Scientific Revolutions,* 2nd ed. (Chicago: University of Chicago Press, 1970), 42.
2. Bruno Latour and Steve Woolgar, *Laboratory Life: The Social Construction of Scientific Facts* (London: Sage, 1979, 252.
3. Mary Hesse, *Revolutions and Reconstructions in the Philosophy of Science* (Bloomington: Indiana University Press, 1980), 188.
4. Ibid., 190.
5. John Ziman, *An Introduction to Science Studies: The Philosophical and Social Aspects of Science and Technology* (Cambridge: Cambridge University Press, 1984), 28.
6. Ibid.
7. See Donald P. Cushman and W. Barnett Pearce, "Generality and Necessity in Three Types of Theory about Human Communication, with Special Attention to Rules Theory," *Human Communication Research* 3 (1977): 344–53.
8. See Ziman's discussion in *Introduction,* 24–28.
9. As a case in point see Lessl's rhetorical study of Carl Sagan's popular public television series, "Cosmos." Lessl says regarding scientists seeking to popularize science: "For the scientist, the need for identification with a nonscientific audience creates a rhetorical dilemma, because the material of science is intrinsically foreign to the uninitiated layman, and to step outside of science to find common ground is to betray the ethos of the scientific community. In this light, one should not be surprised to find the popularizer of science regarded by his professional peers as an outcast or heretic." Thomas M. Lessl, "Science and the Sacred Cosmos: The Ideological Rhetoric of Carl Sagan," *Quarterly Journal of Speech* 71 (1985): 176.
10. Whenever we seek to differentiate "science" from "nonscience" there will be working ambiguities. In practical rhetorical situations scientists will be likely to choose rhetorical strategies that help them construct "boundaries" favorable to their own research aims and claims. In Gieryn's view, scientists engage in "boundary-work" not for the lofty epistemological reason philosophers often cite (e.g., preservation of scientific truth), but as a rhetorical means of solving practical problems that can block achievement of professional goals. Scientists draw sharp contrasts between themselves and nonscientists to enhance their intellectual status and authority vis-à-vis the "out groups," to secure professional resources and career opportunities, to deny these resources and opportunities to "pseudo-scientists," and to insulate scientific research from political interference. It should be added that researchers with unorthodox aims and claims will typically seek to broaden the boundaries of science, attempting to show that they too are scientists and that their claims should also be taken seriously as reasonable scientific contributions. See Thomas F. Gieryn, "Boundary-work and the Demarcation of Science From Non-Science: Strains and Interests in Professional Ideologies of Scientists," *American Sociological Review* 48 (1983): 781–95.
11. The illustrations were chosen because each involved direct argumentative efforts to challenge or bolster the scientific legitimacy of claims made and

the scientific status of those making the claims. Thus, each brings into clear view some standard themes and criteria for scientifically reasonable, rhetorical purposing.

12. My discussion of experimental parapsychology is based on two sociological analyses. Collins and Pinch identify argumentative strategies used to gain or to deny the scientific legitimacy of parapsychology. Allison examines central points of conflict parapsychologists have encountered when seeking to establish parapsychology as a professional, scientific specialty. I have sought to extract from these studies some representative *topoi* used to argue about the scientific reasonableness of parapsychologists' aims. One hoped-for corollary is that commentators will find rhetorical precepts (like *topoi*) heuristically useful in furthering investigations of the sociology of technical scientific claim-making. Sociologists of scientific knowledge have shown that technical claims are subject to social influences and do not result from merely applying neutral procedural rules for validating those claims. Precepts of rhetorical invention identify features of an argumentative logic of science which, unlike formal logic, is applied to make technical claims seem *socially* and *situationally* reasonable to adjudicating audiences. All quotations are taken from these two sources: H. M. Collins and T. J. Pinch, "The Construction of the Paranormal: Nothing Unscientific Is Happening," in Roy Wallis, ed., *On the Margins of Science: The Social Construction of Rejected Knowledge,* Sociological Review Monograph 27 (Staffordshire: University of Keele, 1979), 237–70; and Paul D. Allison, "Experimental Parapsychology as a Rejected Science," in ibid., 271–91.
13. Collins and Pinch, "Construction of the Paranormal," 246–47.
14. E. G. Boring, "Paranormal Phenomena: Evidence, Specification, and Chance," in the introduction to C. E. M. Hansel, *E.S.P.: A Scientific Evaluation* (New York: Scribner's, 1966). Quoted in Collins and Pinch, 246.
15. T. S. Szasz, "A Critical Analysis of the Fundamental Concepts of Psychical Research," *Psychiatric Quarterly* 31 (1957): 96–108. Quoted in Collins and Pinch, 246.
16. J. Beloff, *New Directions in Parapsychology* (London: Elek Science, 1974). Quoted in Collins and Pinch, 247.
17. Collins and Pinch, 250.
18. G. R. Price, "Science and the Supernatural," *Science* 122 (1955): 359–67. Quoted in Collins and Pinch, 250.
19. Ibid. Quoted in Collins and Pinch, 247.
20. D. H. Rawcliffe, *Illusions and Delusions of the Supernatural and the Occult* (New York: Dover, 1959). Quoted in Collins and Pinch, 246.
21. Szasz, "Critical Analysis." Quoted in Collins and Pinch, 248–49.
22. Collins and Pinch, 249.
23. J. C. Crumbaugh, "A Scientific Critique of Parapsychology," *International Journal of Neuropsychiatry* 2 (1966): 539–55. Quoted in Collins and Pinch, 249.
24. D. Cohen, "E.S.P. Science or Delusion?" *The Nation,* 9 May 1966, 550–53. Quoted in Collins and Pinch, 249.
25. Allison, "Experimental Parapsychology," 282.
26. Collins and Pinch (p. 245) draw this quotation from C. Burt, "Psychology

and Parapsychology" in J. R. Smythies, ed., *Science and E.S.P.* (London: Routledge and Kegan Paul, 1967), 62.

27. Collins and Pinch, 242–44; Allison, 279–80.
28. Allison observes that arguments based on this *topos* have harmed parapsychologists' efforts to secure scientific legitimacy; see pp. 281–83.
29. Gertrude Schmeidler and R. A. McConnell, *E.S.P. and Personality Patterns* (New Haven: Yale University Press, 1968). Quoted in Allison, 282.
30. Collins and Pinch (p. 257) draw this quotation from C. Honorton, M. Ramsey and C. Cabibbo, "Experimenter Effects in Extrasensory Perception," *Journal of the American Society of Psychical Research* 69 (1975): 144.
31. Collins and Pinch detail parapsychologists' difficulties in securing access to orthodox journals; see 238, 257–59.
32. Examples used in this discussion of funding are drawn from Allison, 283–84.
33. I have found it necessary to frame this line of thought in a more encompassing manner than originally suggested by Merton's discussion of intellectual communism and similar subsequent uses of the term "communality." Those treatments center attention on the expectation that scientists will share research claims with other members of the community rather than keep them secret for private advantage. Being open with results is not the only way to display communality. Arguments against the parapsychologists suggest that publishing in orthodox journals and securing research funds through recognized public sources can also be used to show active involvement in a scientific community's intellectual life. Failure to display communal participation in those or other ways can not only make a rhetor's place in a scientific community ambiguous and thus diminish perceptions of the rhetor's credibility, but also reduce the likelihood that aims and claims proferred by the rhetor will be accepted as scientifically reasonable. All subsequent uses of communality will signify this more expansive line of thought.
34. Allison, 284–85.
35. J. B. Rhine, "Psi and Psychology: Conflict and Solution," *Journal of Parapsychology* 32 (1968): 118. Quoted in Allison, 285.
36. Bernard Barber, *Science and the Social Order* (1952; rpt. Westport, CT: Greenwood, 1978), 92.
37. Norman W. Storer, *The Social System of Science* (New York: Holt, 1966), 79.
38. Allison, 285. Also see W. O. Hagstrom, *The Scientific Community* (New York: Basic Books, 1965), 211–15.
39. Allison, 285–86.
40. R. A McConnell, *E.S.P.: Curriculum Guide* (New York: Simon and Schuster, 1971), 84. Quoted in Allison, 286.
41. Francine Patterson and Eugene Linden, *The Education of Koko* (New York: Holt, Rinehart, 1981); Thomas A. Sebeok, "The Not So Sedulous Ape," rev. of *The Education of Koko* by Francine Patterson and Eugene Linden, *Times Literary Supplement,* 10 Sept. 1982, 976. Unless otherwise indicated, all quotations from Sebeok's rhetoric are drawn from this review. The

analysis of Sebeok's and Patterson's rhetoric of science is drawn partly from my "The Rhetorical Construction of Scientific Ethos," in Herbert W. Simons, ed., *Rhetoric In the Human Sciences* (London: Sage, 1989), 48–68. This essay extrapolates more fully Sebeok's and Patterson's uses of *topoi* to establish perspectives on scientific ethos.

42. Sebeok's appeal to Terrace's study has added rhetorical force because Terrace had initially believed that his chimpanzee, Nim Chimpsky, was capable of creating a sentence. After further analysis of his data Terrace concluded that Nim said little on his own and was merely imitating behavior in an effort to get rewards; Nim was not able to construct sentences. This study, perhaps more than any other, turned the tide of scientific opinion against the claim that primates could acquire and creatively use langauge. Sebeok used the "reformed" Terrace as persuasive means for questioning the legitimacy of Patterson's claims and the scientific authenticity of her motives. Terrace explained his changed views in "Signs of the Apes, Songs of the Whales" (film produced by Linda Harrar, Nova, WGBH Boston, PBS, 1983).
43. Historically, scientific arguments from the *topos* of skepticism have stressed the positive value of objectivity. That pursuit of objectivity can have negative effects on scientific observation is a relatively rare theme.
44. These sources presumably include Patterson's Gorilla Foundation. In her book she makes a tacit but obvious appeal for financial support of that foundation (*Education of Koko,* 213).
45. Francine Patterson, "Conversation with a Gorilla," *National Geographic,* Oct. 1978, 438–65; condensed in *Reader's Digest,* Mar. 1979, 81–86.
46. Quoted in Cynthia Gorney, "Gorilla Koko Hasn't Convinced Everybody that She Can Talk," *Houston Chronicle,* 4 Feb. 1985, sec. 5, p. 5.
47. Additional *topoi* Patterson uses to construct favorable perceptions of her ethos as a scientist are extrapolated and analyzed in my "Rhetorical Construction of Scientific Ethos."
48. *Education of Koko,* 24–26.
49. Individuality illuminates Patterson's hoped-for ethos better than solitariness, the purported counternorm of Merton's intellectual communism. Rather than using "secrecy" as a scientific virtue, she instead displayed her unwillingness to sacrifice truth to the authority of tradition and its prominent spokespersons. When anti-authoritarianism is treated as a praiseworthy scientific quality it is called "individualism," as Mertonian commentators have noted. This *topos* recommends a line of thought contrary to that suggested by the communality *topos,* as that *topos* was formulated in note 33. For discussion of individuality as a scientific theme see Barber, *Science and the Social Order,* 89–90.
50. *Education of Koko,* 78.
51. Ibid., 79.
52. Ibid., 194.
53. Ibid., 79.
54. Ibid., 80–81.
55. Ibid., 82.
56. Ibid.

57. Ibid., 210–11.

CHAPTER 8

1. Howard Gardner, *Frames of Mind* (New York: Basic Books, 1983), 59–60.
2. This class of exigences is illustrated by Ron Westrum's work on how social factors constrain scientific communities' acceptance or rejection of controversial anomaly claims. See "Science and Social Intelligence About Anomalies: The Case of Meteorites," *Social Studies of Science* 8 (1978): 461–93; "Social Intelligence About Anomalies: The Case of UFO's," *Social Studies of Science* 7 (1977): 271–303; and "Knowledge About Sea-Serpents," in Roy Wallis, ed., *On the Margins of Science: The Social Construction of Rejected Knowledge,* Sociological Review Monograph 27 (Staffordshire: University of Keele, 1979), 293–314.
3. H. M. Collins, "Son of Seven Sexes: The Social Destruction of a Physical Phenomenon," *Social Studies of Science* 11 (1981): 33–62. Collins's study reveals argumentative strategies used to attack and defend claims about the existence of a physical phenomenon. Examination of these arguments shows that subordinate *stases* about evidential and methodological exigences were raised and engaged during the process of the phenomenon's "social destruction." All quotations are from this source.
4. Ibid., 46.
5. Ibid., 40.
6. Westrum, "Social Intelligence about Anomalies," 272.
7. Difficulty in interpreting a set of data can indicate deeper conceptual problems with ordering principles. In this sense definitional *stases* regarding evidential exigences can interact with definitional *stases* about interpretive exigences. But the two kinds of *stases* are nevertheless distinct. As we move a definitional *stasis* from evidential to interpretive matters, the source of ambiguity changes from what data mean to what a given construct means.
8. Stephen Jay Gould, *The Flamingo's Smile: Reflections in Natural History* (New York: Norton, 1985), 247.
9. W. L. Byrne et al., "Memory Transfer," *Science* 153 (1966): 658–59.
10. The actual experimental reports were deposited in the Library of Congress and were available for inspection. These documents reveal that the experiments reported in the letter to *Science* were *not* exact procedural replications of experiments that generated the positive results. This was admitted in some of the experimental reports, but was not mentioned in the letter; Lawrence H. Stern, "Reception of Extraordinary Claims in Science," dissertation in progress.
11. G. D. L. Travis, "Creating Contradiction: Or, Why Let Things Be Difficult When with Just a Little More Effort You Can Make Them Seem Impossible?" Paper presented to the Fifth Annual Meeting of the Society for Social Studies of Science, Toronto, Oct. 1980. Ironically, the authors of the letter urged that scientists *not* make this qualitative judgment. They maintained that neither judgments for nor against the existence of memory transfer were warranted:

    We feel that it would be unfortunate if these negative findings were to be taken as a signal for abandoning the pursuit of a result of enormous

> potential significance. This is especially so in the light of several other related but not identical experiments [they cite relevant literature in an endnote] that support the possibility of transfer of learning by injection of brain-extract from trained donors. Failure to reproduce results is not, after all, unusual in the early phase of research when all relevant variables are as yet unspecified [they cite an article that examines some of these factors] (Byrne et al., "Memory Transfer," 658).

Nevertheless, the letter was generally viewed as a blanket refutation of the transfer phenomenon. This, at least in part, was due to the rhetorical reason that the letter's authors enjoyed strong rhetorical ethos with interested audiences.

12. This class of exigences is illustrated in Andrew Pickering's investigation of social processes involved in scientific theory choice; see "The Role of Interests in High-Energy Physics: The Choice Between Charm and Colour," in Karin D. Knorr, Roger Krohn, and Richard Whitley, eds., *The Social Process of Scientific Investigation,* Sociology of the Sciences Yearbook 4 (Dordrecht: Reidel, 1981), 107–38.
13. Gould, *Flamingo's Smile,* 245.
14. Eventually Linnaeus's four species of coots were rearticulated as four distinct genera. For a discussion of Linnaeus's troubles classifying coots see Thomas Conley, "The Linnaean Blues: Thoughts on the Genre Approach," in Herbert W. Simons and Aram A. Aghazarian, eds., *Form, Genre, and the Study of Political Discourse* (Columbia: University of South Carolina Press, 1986), 60–62. Also see W. T. Stearn, "The Background of Linnaeus's Contributions to the Nomenclature and Methods of Systematic Biology," *Systematic Zoology* 8 (1959): esp. 16–20.
15. Derek E. G. Briggs, Euan N. K. Clarkson, and Richard J. Aldridge, "The Conondont Animal," *Lethaia* 16 (1983): 1–14. Also see Gould's commentary in *Flamingo's Smile,* 249–53.
16. Stebbins and Ayala argued that the theory of punctuated equilibria is weak partly because its authors failed to appreciate these varying criteria while constructing their theory; G. Ledyard Stebbins and Francisco J. Ayala, "Is a New Evolutionary Synthesis Necessary?" *Science* 213 (1981): 968. For a fuller rhetorical analysis of how this controversy ensued before different audiences see John Lyne and Henry F. Howe, "'Punctuated Equilibria': Rhetorical Dynamics of a Scientific Controversy," *Quarterly Journal of Speech* 72 (1986): 132–47.
17. P. M. Driver, "Discussion," in W. C. Corning and S. C. Ratner, eds., *Chemistry of Learning: Invertebrate Research* (New York: Plenum Press, 1967), 61.
18. D. Jensen, in ibid., 59.
19. Stebbins and Ayala's critique of the "new evolutionary synthesis" further illustrates how scientists can take qualitative "stands" on problems of interpretation. They framed their case with three questions:

> (i) whether microevolutionary processes *operate* (and have operated in the past) throughout the different taxa in which macroevolutionary phenomena are observed; (ii) whether the microevolutionary processes identified by population geneticists (mutation, chromosomal change, random

> drift, natural selection) can account for the morphological changes and other macroevolutionary phenomena observed in higher taxa . . . ; and (iii) whether evolutionary trends and other macroevolutionary patterns can be predicted from knowledge of microevolutionary processes ("Is a New Evolutionary Synthesis Necessary?" 968).

These three questions pose qualitative *stases*. Stebbins and Ayala address these issues on pp. 968–71.

20. Thomas S. Kuhn, "Objectivity, Value Judgment and Theory Choice," in his *The Essential Tension: Selected Studies in Scientific Tradition and Change* (Chicago: University of Chicago Press, 1977), 323.
21. Thomas S. Kuhn, "Reflections on My Critics," in Imre Lakatos and Alan Musgrave, eds., *Criticism and the Growth of Knowledge* (London: Cambridge University Press, 1970), 262.
22. This point is discussed in G. Nigel Gilbert, "The Transformation of Research Findings into Scientific Knowledge," *Social Studies of Science* 6 (1976): 285–87.
23. Collins's analysis of the gravitational radiation controversy as involving "negotiations" about what constitutes a "working" gravitational wave detector illustrates scientists attempting to modify methodological ambiguities or exigences; see H. M. Collins, "The Seven Sexes: A Study in the Sociology of a Phenomenon, or the Replication of Experiments in Physics," *Sociology* 9 (1975): 205–24. Also see his "Son of Seven Sexes."
24. For explanation of this practice see Henry Harris, "Rationality in Science," in A. F. Heath, ed., *Scientific Explanation* (Oxford: Clarendon Press, 1981), 44–45.
25. Francine Patterson and Eugene Linden, *The Education of Koko* (New York: Holt, Rinehart, 1981), 78–79.
26. For a useful study of a controversy about the best taxonomic methods see John Dean, "Controversy over Classification: A Case Study from the History of Botany," in Barry Barnes and Steven Shapin, eds., *Natural Order: Historical Studies of Scientific Culture* (London: Sage, 1979), 211–30.
27. For one effort to resolve this issue see Ernst Mayr, "Biological Classification: Toward a Synthesis of Opposing Methodologies," in Elliott Sober, ed., *Conceptual Issues in Evolutionary Biology* (Cambridge, MA: MIT Press, 1984), 646–62.
28. James V. McConnell, "The Biochemistry of Memory," in Corning and Ratner, *Chemistry of Learning,* 310–22. The paper was prepared initially as an address to an immediate audience. McConnell's adaptations included formulating controversial issues sharply and explicitly and arguing for his stands on those issues with a forceful, personalized style, unlike the more passive inductivist style so typical of journal articles. I chose McConnell's paper for analysis precisely for these reasons. He constructed his paper for presentation at a symposium—a rhetorical situation in which rhetors can anticipate that their claims will be discussed. In such dynamic rhetorical contexts, rhetors are likely to bring controversial issues forward explicitly as personal points for contention.
29. At that time McConnell's aim was to search for the *engram,* a physical representation of memory thought to be located at the synapse between two

nerve cells. McConnell reasoned that because worms are the lowest form of life that have true synapses, they should be capable of learning. He sought to test the physical hypothesis with this experiment. Only through hindsight, after biochemical theories of memory became important, did this early experiment appear relevant to investigation of the biochemical hypothesis. For this and other points pertaining to the social history of the memory-transfer controversy I have relied on Stern, "Reception of Extraordinary Claims in Science."

30. McConnell describes the first regeneration experiment in "The Biochemistry of Memory," 310–11.
31. Ibid., 312–13.
32. The three classes of refutative arguments are elaborated in G. D. L. Travis, "Replicating Replication? Aspects of the Social Construction of Learning in Planarian Worms," *Social Studies of Science* 11 (1981): 16–25.
33. A rhetorically significant example is a paper by Edward L. Bennett and Nobel chemist Melvin Calvin in which they claimed worms could not be trained reliably. Some believed this article, coupled with the powerful ethos of Calvin, slowed interest in planarian research. As W. C. Corning, a psychologist who conducted and supported planarian transfer research said: "Now if Bennett and Calvin meant to say that worms could not be trained *reliably,* this message did not get across. I, too, am frequently confronted with the statement, 'Calvin says worms can't learn,' and with a Nobel, people tend to believe this. However, I think that he is a better chemist than psychologist"; "Discussion," in Corning and Ratner, *Chemistry of Learning,* 273. Also see Edward L. Bennett and Melvin Calvin, "Failure to Train Planarians Reliably," *Neuroscience Research Program Bulletin* 2 (1964): 3–24.
34. The explanations I offer for sensitization, pseudoconditioning, and learning should not obscure the situationally relevant point that definitions and experimental applications were debatable throughout the planarian controversy.
35. Two important studies using sensitization to challenge specific memory-transfer explanations are A. L. Hartry, P. Keith-Lee, and W. D. Morton, "Planaria: Memory Transfer Through Cannibalism Re-Examined," *Science* 146 (1964): 274–75; and D. R. Walker and G. A. Milton, "Memory Transfer vs. Sensitization in Cannibal Planarians," *Psychonomic Science* 5 (1966): 293–94.
36. For an example of this line of criticism see Donald D. Jensen, "Paramecia, Planaria, and Pseudo-Learning," *Animal Behaviour* 13 Suppl. 1 (1965): 9–20.
37. "Replicating Replication," 19, 22–25.
38. McConnell's scientific ethos was diminished in the eyes of some scientists due to his (1) speculations in the popular press about the practical applications of chemical memory transfer, and (2) scientifically unorthodox sense of humor and willingness to present the "backstage" happenings of science. His unorthodoxy contributed to perceptions that he was not sufficiently "earnest" in scientific matters. On this point see David Travis, "On the Construction of Creativity: The 'Memory Transfer' Phenomenon and the

Importance of Being Earnest," in Knorr, et al., *Social Process of Scientific Investigation,* esp. 173–78. Also see Stern, "Reception of Extraordinary Claims."

39. "Biochemistry of Memory," 310.
40. Ibid., 311.
41. Ibid., 313.
42. Ibid., 312.
43. Ibid.
44. Ibid., 313.
45. Ibid., 314.
46. Ibid.
47. Ibid.
48. Ibid.
49. Ibid., 316.
50. McConnell's rhetorical subtlety comes into full view only when some missing parts of the context are filled in. Of the three authors, Hartry and Keith-Lee were still graduate students and Morton was an undergraduate; none was an "accredited" scientist, and McConnell thought that this showed. Furthermore, the three students were not attending a prestigious research institution. McConnell chose to challenge the experiment on technical grounds rather than attack the three scientists' credibility. I am indebted to Stern for this insight.
51. "Biochemistry of Memory," 316–17.
52. Ibid., 318.
53. Ibid.
54. Ibid.
55. Ibid., 318–19.
56. Ibid., 319.
57. Ibid., 320.
58. Ibid.
59. Ibid.
60. Ibid.
61. Ibid., 320–21.
62. Wells, "Discussion," in Corning and Ratner, *Chemistry of Learning,* 324.
63. Ibid.
64. Ibid.
65. Bennett, "Discussion," in Corning and Ratner, *Chemistry of Learning,* 325.
66. For extended discussion of this problem see M. Ray Denny, "A Learning Model," in Corning and Ratner, *Chemistry of Learning,* 32–42; John Gaito, "The Possible Role of RNA in Learning and Memory Events," ibid., 23–31; and comments about those papers in "Discussion," ibid., 56–63.
67. McConnell's rejection of semantic problems and the points made in his final evaluative appeal become fully meaningful within the orientational context of his effort to forge a chemical theory of learning. See James V. McConnell, "Cannibals, Chemicals, and Contiguity," *Animal Behaviour* 13 Suppl. 1 (1965): 61–66.
68. Collins, "Seven Sexes," 206–07.

69. Travis, "Replicating Replication," esp. 12–13, 22–25.
70. McConnell, "Specific Factors Influencing Planarian Behavior," in Corning and Ratner, *Chemistry of Learning,* 217–33.
71. Ibid., 217.
72. Ibid., 231.
73. E. Halas, "Discussion," in Corning and Ratner, *Chemistry of Learning,* 274.
74. Ibid., 273.
75. Corning, "Discussion," in Corning and Ratner, *Chemistry of Learning,* 274.
76. Bennett, "Discussion," in Corning and Ratner, *Chemistry of Learning,* 272.
77. For instance, see W. C. Corning and D. Riccio, "The Planarian Controversy," in W. L. Byrne, ed., *Molecular Approaches to Learning and Memory* (New York: Academic Press, 1970), 128.
78. As will be shown in chapter 10, similar conjectural challenges were made against the claim that creationism is science.
79. M. L. Clarke thought the emphasis of Roman education on rhetoric in general and invention in particular stifled independent and creative thinking, encouraging "a certain conventionality of thought." Ehninger thought that *stasis* doctrine, along with other classical tools of invention—*topoi* and *loci*—were "artificial substitutes for knowledge and cognition." See M. L. Clarke, *Rhetoric at Rome* (New York: Barnes and Noble, 1963), 158–64; Douglas Ehninger, "On Systems of Rhetoric," *Philosophy and Rhetoric* 1 (1968): 134; and his "George Campbell and the Revolution in Inventional Theory," *Southern Speech Communication Journal* 25 (1950): 270.

CHAPTER 9

1. In rhetorical terms, "the social negotiation of reality" involves inspecting *topoi* for situationally reasonable notions for urging acceptance of claims. This is not inconsistent with Harvey's discussion of the role of plausibility in evaluating knowledge claims. The term "plausibility" is used to make the point that scientists appraise claims for their acceptability within the constraints of a social context and not for their formally logical character or intrinsic empirical truth. Harvey's rich case study supports this view. I would amend that case by observing that it also reveals the functioning of experimental competence and other rhetorical *topoi* as the bases of rhetorical strategies scientists used publicly to justify their appraisals; see Bill Harvey, "Plausibility and the Evaluation of Knowledge: A Case Study of Experimental Quantum Mechanics," *Social Studies of Science,* 11 (1981), 95–130.
2. This discussion is based on ibid., esp. 107–15, and Bill Harvey, "The Effects of Social Context on the Process of Scientific Investigation: Experimental Tests of Quantum Mechanics," in Karin D. Knorr, Roger Krohn, and Richard Whitley, eds., *The Social Process of Scientific Investigation,* Sociology of the Sciences Yearbook 4 (Dordrecht: Reidel, 1981), 139–63, esp. 148–52.
3. Quoted in Harvey, "Effects of Social Context," 149.

4. Ibid., 150–51.
5. See Boyce Rensberger, "Margaret Mead: The Nature-Nurture Debate I," *Science 83* 4 (April 1983): 28–37, for discussion of Derek Freeman's *Margaret Mead and Samoa: The Making and Unmaking of an Anthropological Myth* (Cambridge, MA: Harvard University Press, 1983).
6. H. M. Collins, "The Seven Sexes: A Study in the Sociology of a Phenomenon, or the Replication of Experiments in Physics," *Sociology* 9 (1975): 216. Whether this *topos*—like all *topoi*—can be applied persuasively depends on the constraining influences of the rhetorical situation. Features of Collins's enculturation model or the contrasted algorithmic model can be invoked distinctly or in some combination when arguments are constructed on the basis of this *topos*. Persuasiveness will vary with what claims audiences are willing to accept as situationally reasonable.
7. This illustration is drawn from Brian Wynne, "C. G. Barkla and the J Phenomenon: A Case Study in the Treatment of Deviance in Physics," *Social Studies of Science* 6 (1976): 332 (emphasis added). Wynne argues that replicability and other themes are offered by defenders of scientific orthodoxy as after-the-fact rhetorical rationalizations for rejecting unorthodox claims which they in fact dismiss for social and pragmatic reasons. He thus challenges the formally rational model for science and calls for an alternative. Rhetorical reasonableness might be that alternative. Rhetors can persuasively apply the rhetorical logic of science only through choosing *topoi* that fittingly engage the pragmatic and social interests constraining what judging audiences will accept as situationally reasonable.
8. See H. M. Collins, "Son of Seven Sexes: The Social Destruction of a Physical Phenomenon," *Social Studies of Science* 11 (1981): 51.
9. Quoted in Harvey, Plausibility and the Evaluation of Knowledge," 115.
10. For discussion of possible differences see ibid., 114–15, and Harvey, "Effects of Social Context," 159–60.
11. William J. Broad, "The Publishing Game: Getting More for Less: Meet the Least Publishable Unit, One Way of Squeezing More Papers out of a Research Project," *Science* 211 (1981), 1137–39. The Meyers quotation is from this article.
12. Quoted in S. W. Woolgar's study of how scientists differently reconstructed the process of pulsar discovery, "Writing an Intellectual History of Scientific Development: The Use of Discovery Accounts," *Social Studies of Science* 6 (1976): 402.
13. Ibid., 404.
14. A. Hewish, S. J. Bell, J. D. H. Pilkington, P. F. Scott, and R. A. Collins, "Observation of a Rapidly Pulsating Radio Source," in *Pulsating Stars: A Nature Reprint* (London: Macmillan, 1968), 5.
15. Ibid.
16. John Ziman, *Public Knowledge* (London: Cambridge University Press, 1968), 41.
17. Thomas S. Kuhn, "The Historical Structure of Scientific Discovery," in his *The Essential Tension: Selected Studies in Scientific Tradition and Change* (Chicago: University of Chicago Press, 1977), 166–67.

18. Ibid., 175.
19. Thomas S. Kuhn, *The Structure of Scientific Revolutions,* 2nd ed. (Chicago: University of Chicago Press, 1970), 154–55. Also see his *The Copernican Revolution: Planetary Astronomy in the Development of Western Thought* (Cambridge, MA: Harvard University Press, 1957), 219–25.
20. See Ron Westrum, "Knowledge About Sea-Serpents", in Roy Wallis, ed., *On the Margins of Science: The Social Construction of Rejected Knowledge,* Sociological Review Monograph 27 (Staffordshire: University of Keele, 1979), 307.
21. *Structure of Scientific Revolutions,* 154.
22. Bernard Lovell, *Out of the Zenith* (New York: Harper & Row, 1973), 130–31.
23. These propositions are based on Kuhn's discussion of three sorts of puzzles scientists engage in basic, normal research. See "The Essential Tension: Tradition and Innovation in Scientific Research," in *Essential Tension,* 233.
24. Wynne, "C. G. Barkla and The J Phenomenon," 337.
25. Kuhn observes an anomaly "must usually be more than an anomaly" to evoke crisis. The point could be added that arguments must be made establishing an anomaly as significant for a scientific community. Such arguments are based on the significant anomaly *topos,* regardless of whether they do or do not culminate ultimately in intellectual crisis for the community. See *Structure of Scientific Revolutions,* 81–82.
26. "Historical Structure of Scientific Discovery," 171.
27. Quoted in Woolgar, "Writing an Intellectual History," 402.
28. Laurance N. Fredrick and Robert H. Baker, *An Introduction to Astronomy,* 8th ed. (New York: Van Nostrand, 1974), 340–41.
29. T. J. Pinch is specially interested in understanding how "investments of credibility" undertaken by both experimenters and theoreticians constrained reception of Davis's result, leading ultimately to establishment of the result as credible despite its clash with received theory; see "Theoreticians and the Production of Experimental Anomaly: The Case of Solar Neutrinos," in Knorr, et al., *Social Process of Scientific Investigation,* esp. 92–102.
30. Ibid., 93–94, 95, 100.
31. Quoted in ibid., 99.
32. Ibid., 99–100.
33. *Structure of Scientific Revolutions,* 153. Similarly, Brannigan claims that research is granted the status of scientific discovery in part because "the announcement is coherent with existing knowledge in the field or that it *resolves certain outstanding problems,* as for example the discovery of Neptune by Leverrier explained the perturbations of Uranus, or the discovery of the bihelical structure of DNA by Crick and Watson explained the symmetry of chromosomal splitting and the templating of RNA from the DNA molecule" (emphasis added). What Brannigan could have pointed out is that discovery claims are strengthened whenever a convincing line of argument can be developed linking those claims with solution of problems

that specially concern a scientific community. See Augustine Brannigan, *The Social Basis of Scientific Discoveries* (Cambridge: Cambridge University Press, 1981), 72.

34. For instance, see Thomas S. Kuhn, "Objectivity, Value Judgment, and Theory Choice," *Essential Tension*, 321–24; *Structure of Scientific Revolutions*, 184–86; and, for elegance, see ibid., 155–56.
35. Kuhn, "Objectivity, Value Judgment, and Theory Choice," 323.
36. W. V. Quine and J. S. Ullian, *The Web of Belief*, 2nd ed. (New York: Random House, 1978), 66–68.
37. Richard Whately, *Elements of Rhetoric*, Douglas Ehninger, ed. (Carbondale: Southern Illinois University Press, 1963), 112.
38. Ibid., 115.
39. David Travis, "On the Construction of Creativity: The 'Memory Transfer' Phenomenon and the Importance of Being Earnest," in Knorr, et al., *Social Process of Scientific Investigation*, 169.
40. Jean Umiker-Sebeok and Thomas A. Sebeok, "Introduction: Questioning Apes," in Umiker-Sebeok and Sebeok, eds., *Speaking of Apes: A Critical Anthology of Two-Way Communication with Man* (New York: Plenum Press, 1980), 14. Several illustrations of arguments from the *topos* of parsimony can be found on pp. 14–21.
41. Ibid., 14–15.
42. Ibid., 15–16.
43. Ibid., 15 n.
44. Donald R. Griffin, "Experimental Cognitive Ethology," *The Behavioral and Brain Sciences* 1 (1978): 555.
45. Bertrand Russell, *Mysticism and Logic* (New York: Longmans, Green, 1925), 60.
46. Steven Rose, *The Conscious Brain* (1973; rpt. New York: Knopf 1976), 207 (emphasis added).
47. Kuhn, "Objectivity, Value Judgment, and Theory Choice," 322 n.
48. Lovell, *Out of the Zenith*, 135.
49. Chaim Perelman and L. Olbrechts-Tyteca, *The New Rhetoric* (Notre Dame, IN: University of Notre Dame Press, 1969), part 3, ch. 3.
50. Determining the role of metaphor and analogy in scientific theory and investigation is an immensely complex problem in the philosophy of science. I do not deal with those complexities here because my primary purpose is to determine how metaphors, analogies, and examples function as exemplary rhetorical *topoi*. More philosophically oriented discussions can be found in such works as Max Black, *Models and Metaphors* (Ithaca, NY: Cornell University Press, 1962); Mary Hesse, *Revolutions and Reconstructions in the Philosophy of Science* (Bloomington: Indiana University Press, 1980), esp. ch. 4, and her *Models and Analogies in Science* (Notre Dame, IN: University of Notre Dame Press, 1966); and Andrew Ortony, ed., *Metaphor and Thought* (Cambridge: Cambridge University Press, 1979).
51. Black, *Models and Metaphors*, 227. Maxwell's quotations also are found on this page.
52. Ibid., 228.

53. Andrew Pickering, "The Role of Interests in High-Energy Physics: The Choice Between Charm and Colour," in Knorr, et al., *Social Process of Scientific Investigation,* 109.
54. Lewis Thomas, *The Lives of a Cell* (New York: Bantam Books, 1974), 18.
55. Perelman and Olbrechts-Tyteca, *New Rhetoric,* 350.
56. Ibid., 357.
57. Ibid., 351.
58. Ibid., 372.
59. Arthur Koestler referred disparagingly to this practice as the "ratomorphic" fallacy of behavioral psychology; see *The Ghost in the Machine* (Chicago: Regnery, 1967), 17.
60. Discussions of limited traditional views and articulations of more expansive notions of metaphor are found in I. A. Richards, *The Philosophy of Rhetoric* (1936; rpt. London: Oxford University Press, 1976), 89–112; and Black, *Models and Metaphors,* 25–47.
61. *Philosophy of Rhetoric,* 94.
62. *Models and Metaphors,* 236–37. In Black's view metaphors and scientific models are similar insofar as their respective uses depend on analogical thought processes; he distinguishes metaphors from models on the grounds that metaphors are based on commonplace associations or "proverbial knowledge" while models require comprehension of a "well-knit theory" (238–39). Black takes this line of thought even further when he discusses an "implicit or submerged model" called an *archetype.* An archetype is a "systematic repertoire of ideas by which a given thinker describes, by *analogical extension,* some domain to which those ideas do not immediately and literally apply" (241).
63. S. Michael Halloran and Annette Norris Bradford, "Figures of Speech in the Rhetoric of Science and Technology," in Robert J. Connors, Lisa S. Ede, and Andrea A. Lunsford, eds., *Essays on Classical Rhetoric and Modern Discourse* (Carbondale: Southern Illinois University Press, 1984), 184.
64. Ibid., 186. For full discussion of this metaphorical system see 183–88.
65. Perelman and Obrechts-Tyteca, *New Rhetoric,* 405–10. This is precisely how Brown accounts for the conflict between evolutionary functionalism and experimental empiricism in historical sociology; Richard Harvey Brown, "Rhetoric and the Science of History: The Debate Between Evolutionism and Empiricism as a Conflict of Metaphors," *Quarterly Journal of Speech* 72 (1986): 148–61, esp. 149, 152.
66. John C. Marshall, "Minds, Machines, and Metaphors," *Social Studies of Science* 7 (1977): 475. Other examples of metaphorical clashes are revealed in an excellent historical survey of various machine metaphors used to account for human brain functioning (475–88).
67. Umiker-Sebeok and Sebeok, *Speaking of Apes,* 16.
68. Historical content is drawn primarily from the following sources: Edwin Clarke and C. D. O'Malley, *The Human Brain and Spinal Cord: A Historical Study Illustrated by Writings from Antiquity to the Twentieth Century* (Berkeley: University of California Press, 1968); Aleksandr Romanovich Luria, *Higher Cortical Functions in Man,* 2nd ed. (New York: Basic Books, 1980); Jürgen Thorwald, *The Triumph of Surgery,* R. W. Winston and C.

Winston, trans. (New York: Pantheon, 1960), 3–40; A. Earl Walker, "The Development of the Concept of Cerebral Localization in the Nineteenth Century," *Bulletin of the History of Medicine* 31 (1957): 99–121; Robert M. Young, *Mind, Brain and Adaptation in the Nineteenth Century: Cerebral Localization and Its Biological Context from Gall to Ferrier* (Oxford: Clarendon Press, 1970). Useful preliminary ideas about exemplary arguments and relevant historical content are found in Susan Leigh Star, "Tactics in the Debate About Cerebral Localization, 1870–1905," paper presented to the Society for the Social Studies of Science, Oct. 1982, Philadelphia.

69. A. Earl Walker, "Stimulation and Ablation: Their Role in the History of Cerebral Physiology," *Journal of Neurophysiology* 20 (1957): 435–49.
70. See David Ferrier, "Discussion on the Localization of Function in the Cortex Cerebri," in Robert H. Wilkins, ed., *Neurosurgical Classics* (New York: Johnson Reprint Corporation, 1965), 125–28.
71. Ibid., 128.
72. Discussion of Goltz's position is based on Luria, *Higher Cortical Functions,* 13–14; Thorwald, *Triumph of Surgery,* 24–29, 33–35; and Clarke and O'Malley, *The Human Brain,* 558–65.
73. See Thorwald, *Triumph of Surgery,* 38–39.
74. David Ferrier, *The Functions of the Brain* (Smith, Elder, & Co., 1876), 305–7. Compare the points on Fig. 64, which shows a lateral view of a monkey brain, with those on Fig. 63, which presents a human brain from the same perspective (304). Also compare the points as they appear on Fig. 66 and Fig. 65, which show respectively the surface of a monkey and a human brain. Fig. 64 and Fig. 63 also are available in Edwin Clarke and Kenneth Dewhurst, *An Illustrated History of Brain Function,* (Berkeley: University of California Press, 1972), 114–15.
75. Star called this sort of argument an "argument from classification," treating it as a species of argument by example; "Tactics," 4.
76. Ferrier, *Functions of the Brain,* 305.
77. Ibid., 296–97.
78. Ferrier, "Discussion on the Localization of Function," 125.
79. Thorwald, *Triumph of Surgery,* 39.
80. Clarke and Dewhurst, *Illustrated History of Brain Function,* 115; Star, "Tactics," 8–10; Thorwald, *Triumph of Surgery,* 12–13; and Young, *Mind, Brain and Adaptation,* 243–45.
81. Star acknowledges the importance of mapping arguments, but does not analyze them as metaphors; see "Tactics," 4.
82. Walter Campbell, *Histological Studies on the Localisation of Cerebral Function* (Cambridge: Cambridge University Press, 1905), 292–93.

CHAPTER 10

1. In this sense rhetoric becomes the central agency of change within the framework of Feyerabend's epistemological anarchism. Scheffler sees a similar notion of rhetoric as something that must be guarded against to preserve science as a reasonable, responsible, principled endeavor. See Paul

Feyerabend, *Against Method* (1975; rpt. London: Verso, 1978); and Israel Scheffler, *Science and Subjectivity* (Indianapolis: Bobbs-Merrill, 1967).

2. Sources in which these other kinds of investigations are attempted and presented include John S. Nelson, Allen Megill, and Donald N. McCloskey, eds., *The Rhetoric of the Human Sciences: Argument in Scholarship and Public Affairs* (Madison: University of Wisconsin Press, 1987); S. Michael Halloran and Annette Norris Bradford, "Figures of Speech in the Rhetoric of Science and Technology," in Robert J. Connors, Lisa S. Ede, and Andrea A. Lunsford, eds., *Essays on Classical Rhetoric and Modern Discourse* (Carbondale: Southern Illinois University Press, 1984), 179–92; Alan G. Gross, "Style and Arrangement in Scientific Prose: The Rules Behind the Rules," *Journal of Technical Writing and Communication*, 14 (1984): 241–53; and "The Form of the Experimental Paper: A Realization of the Myth of Induction," *Journal of Technical Writing and Communication*, 15 (1985): 15–26.
3. For an overview of the trial's context see La Follette's useful introduction to Marcel C. La Follette, ed., *Creationism, Science, and the Law: The Arkansas Case* (Cambridge, MA: MIT Press, 1983). This volume provides an excellent selection of materials on the *McLean v. Arkansas* trial. The text of Act 590, excerpts from both the plaintiffs' and the defendants' preliminary outlines and pretrial briefs, and Overton's opinion are among these materials. Also included are articles by attorneys involved with the McLean case, academic witnesses for the plaintiffs, and science advisers to the plaintiffs. All references to Overton's opinion will be to the text presented in *McLean v. Arkansas*, United States District Court (Eastern District, Arkansas, 1982), vol. 529 Federal Supplement, 1255–74.
4. Quoted in *McLean*, 1258.
5. Ibid., 1257.
6. Makau develops the point that the Supreme Court must balance social, political, and economic demands with principles establishing doctrinal consistency when rendering reasonable decisions. Overton appears to be seeking such balance in his opinion. See Josina M. Makau, "The Supreme Court and Reasonableness," *Quarterly Journal of Speech* 70 (1984): 379–96.
7. La Follette summarizes the plaintiffs' case in the introduction to *Creationism, Science and the Law*, 7–8.
8. *McLean*, 1264.
9. Ibid., 1272.
10. Ibid.
11. Thomas F. Gieryn, George M. Bevins, and Stephen C. Zehr, "Professionalization of American Scientists: Public Science in the Creation/Evolution Trials," *American Sociological Review* 50 (1985): 393. According to the authors, the type of professional boundary work used in *McLean* tries "to protect public investments in science by excluding as non-scientific a potential competitor for those resources" (394). Creationism was depicted as pseudoscientific religion in direct clash with genuine science. There was no sense of complementarity among distinct kinds of truth. Instead, there was direct clash about what counts as scientific truth and falsehood (406).

12. Eric Holtzman and David Klasfeld, "The Arkansas Creationism Trial: An Overview of the Legal and Scientific Issues," La Follette, *Creationism, Science, and the Law,* 94. Holtzman was Professor and Chair of the Department of Biological Sciences at Columbia University and science adviser to the American Civil Liberties Union; Klasfeld was a trial attorney for the plaintiffs.
13. Holtzman and Klasfeld observe that "deeper" questions like these were not addressed with self-indulgent, complex philosophical arguments. My point is that they *were* addressed, but at a level of generality that was rhetorically reasonable for an audience inexpert in philosophy or science. I do not consider these two views incompatible. See ibid, 93–94.
14. Gieryn et al., "Professionalization," 403. The authors make clear that they are interpreting Mertonian norms as "ideological statements used in *public* science to advance narrow legal goals and broader professional goals at stake in *McLean*" (403). I am interpreting them as rhetorical *topoi* for constructing and evaluating scientific ethos.
15. Ibid., 401. The authors also discuss creationists' lack of training in specialized research techniques as another reason for their lack of communality (see 401–2). I shall discuss this argument later in another connection.
16. Ibid., 401.
17. According to Dorothy Nelkin, creationists commonly depicted evolutionary biologists in this way; see "Legislating Creation in Arkansas," *Society* 20 (1983): 15.
18. McLean, 1268.
19. Quoted in Gene Lyons, "Repealing the Enlightenment," in Ashley Montagu, ed., *Science and Creationism* (New York: Oxford University Press, 1984), 358.
20. Gieryn et al., "Professionalization," 402.
21. Michael Ruse, "Creation-Science Is Not Science," in La Follette, *Creationism, Science, and the Law,* p. 157.
22. *McLean,* 1260 n. 7.
23. Gieryn et al., "Professionalization," 402–03.
24. *McLean,* 1269.
25. According to Holtzman and Klasfeld, scientists wanted to center discussion on Wickramasinghe's theories as poor science, but the lawyers thought it a better tack to turn his testimony against creationist postulates. This strategy avoided framing the controversy in terms of comparing "good" and "weak" science. See their full discussion of Wickramasinghe's testimony in "The Arkansas Creationism Trial," 95–96.
26. *McLean,* 1269–70.
27. Gieryn et al., "Professionalization," 403.
28. *McLean,* 1264.
29. Ibid.
30. Ibid., 1261.
31. Quoted in ibid.
32. Quoted in ibid., 1262.
33. Ibid., 1267.

34. Mark E. Herlihy, "Trying Creation: Scientific Disputes and Legal Strategies," in La Follette, *Creationism, Science, and the Law,* 102.
35. Laudan excellently identifies points at issue between scholars who believe science is best characterized by agreement and those who focus instead on scientific controversy. One conclusion that can be drawn from these philosophical and sociological debates is that there is no scholarly consensus concerning answers to such questions as What is science? and Who are "real" scientists? We therefore know that when "experts" enter public forums to offer univocal answers to such questions, they rhetorically construct and promote adherence to a singular but nevertheless disputable vision of science and its practitioners. See Larry Laudan, "Two Puzzles About Science: Reflections About Some Crises in the Philosophy and Sociology of Science," *Minerva* 20 (1982): 253–68.
36. *McLean,* 1264.
37. 1981 Arkansas Act 590, Sec. 4, 1232–33.
38. Larry Laudan, "Commentary on Ruse: Science at the Bar—Causes for Concern," in La Follette, *Creationism, Science, and the Law,* 166.
39. Ibid., 164.
40. Ruse helped to advance five key criteria for science that Overton used to adjudicate *McLean:* (1) it is guided by natural law, (2) it is explanatory with reference to natural law, (3) it is testable, (4) it is tentative, and (5) it is falsifiable. Said Ruse about professing these criteria in *McLean:* "I would like to pride myself that the points I made . . . went straight through into Judge Overton's decision." See Michael Ruse, "A Philosopher's Day in Court," in Montagu, *Science and Creationism,* 322, and *McLean,* 1267. Also see Ruse, "Response to Laudan's Commentary: Pro Judice," in La Follette, *Creationism, Science and the Law,* esp. 169.
41. Joel Cracraft, "The Scientific Response to Creationism," La Follette, *Creationism, Science, and the Law,* 139.
42. Laudan, "Commentary on Ruse," 162.
43. Ibid., 165.
44. Creationists exploit the general public's ignorance of scientific analysis and its reverence for scientific authority by adopting the *persona* of scientists as they seek to induce adherence to political or religious claims. See Gieryn et al., "Professionalization," 403; Gary E. Crawford, "Science as an Apologetic Tool for Biblical Literalists," in La Follette, *Creationism, Science, and the Law,* 111–12; Cracraft, "Scientific Response," ibid., 138–39; La Follette, introduction, ibid.; and Dorothy Nelkin, "From Dayton to Little Rock: Creationism Evolves," ibid., 77. The persuasive power of this adopted scientific *persona* is exemplified when creationists successfully debate orthodox scientists before the laity. For discussions of creationists' rhetorical capacity to confound their opponents through public debate see Roy A. Gallant, "To Hell with Evolution," in Montagu, *Science and Creation,* 300–305; and Gene Lyons, "Repealing the Enlightenment," ibid., 355–57.
45. Holtzman and Klasfeld identified this general creationist strategy, observing that "creationists are adept at public debate in which they stay on the offensive, diverting attention from the fissures in their arguments by high-

lighting each flaw, no matter how minor, in opposing views." See "The Arkansas Creationism Trial," 91.

46. Ibid., 92.
47. "Response to Laudan," 168.
48. *McLean,* 1269.
49. Ibid., 1270.
50. Ibid.
51. Ibid., 1265.
52. Ibid., 1267.
53. Ibid., 1265 n. 19.
54. Ibid., 1267–68.
55. Ibid., 1267.
56. Ibid.
57. Ruse, "Creation-Science Is Not Science," 156.
58. *McLean,* 1268.
59. For a useful survey of technical arguments against creationists' claims see Cracraft, "Scientific Response," 138–49. More focused critiques and refutations of the case for creationism are collected in Laurie R. Godfrey, ed., *Scientists Confront Creationism* (New York: Norton, 1983). Also see Philip Kitcher, *Abusing Science: The Case Against Creationism* (Cambridge, MA: MIT Press, 1982).
60. See Forbes Hill, "Conventional Wisdom—Traditional Form: The President's Message of November 3, 1969," *Quarterly Journal of Speech* 58 (1972): 373–86; Karlyn Kohrs Campbell, " 'Conventional Wisdom—Traditional Form': A Rejoinder," ibid., 451–54; and Hill's, "Reply to Professor Campbell," ibid., 454–60.
61. Horace Freeland Judson, *The Eighth Day of Creation: Makers of the Revolution in Biology* (New York: Simon and Schuster, 1979); Robert Olby, *The Path to the Double Helix* (Seattle: University of Washington Press, 1974). Useful technical background is contained in Garland Allen, *Life Science in the Twentieth Century* (New York: Wiley, 1975), esp. 187–228.
62. See James D. Watson, *The Double Helix* (New York: New American Library, 1968); Anne Sayre, *Rosalind Franklin and DNA* (New York: Norton, 1975).
63. J. D. Watson and F. H. C. Crick, "A Structure for Deoxyribose Nucleic Acid," *Nature* 171 (1953): 737–38. Due to the brevity of this article, references will be to paragraphs by number, 1–14.
64. See Pauling's discussion of his error, "Molecular Basis of Biological Specificity," *Nature* 248 (1974): 771.
65. For instance, Jerry Donahue helped Watson and Crick get the form of the bases right and John Griffith influenced their thought about the complementarity of the base-pairing mechanism. For a list of the rhetors, both major and minor, see Judson, *Eighth Day,* 24–25.
66. See Rosalind E. Franklin and R. G. Gosling, "Molecular Configuration in Sodium Thymonucleate," *Nature* 171 (1953): 740–41; M. H. F. Wilkins, A. R. Stokes, and H. R. Wilson, "Molecular Structure of Deoxypentose Nucleic Acid," *Nature* 171 (1953): 738–40.
67. For a study of how topical and stylistic choices in the article created a

Watson-Crick ethos with potentially revolutionary implications for other scientists, see S. Michael Halloran, "The Birth of Molecular Biology: An Essay in the Rhetorical Criticism of Scientific Discourse," *Rhetoric Review* 3 (1984): 70–83. For close analysis of the article's language see Charles Bazerman, "What Written Language Does: Three Examples of Academic Discourse," *Philosophy of the Social Science.* 11 (1981): 364–69. Both articles are useful in amplifying features of the rhetorical logic of science reflected in Watson and Crick's paper, although this was not the explicit aim of either author.

68. Judson explains principles supporting these reasons for rejection in *Eighth Day,* 79–80, 158. We learn from Watson that he and Crick thought Pauling's neglect of binding forces was an incredible theoretical blunder (see *Double Helix,* 102–03). In the language of *stasis* theory, Pauling failed to make *quality* applications of received chemical principles in constructing his model. Franklin had methodological and evidential reasons for rejecting the model which, probably, were not known initially by Watson and Crick. Pauling and Corey had used Astbury's x-ray data, which Franklin knew were spurious because laboratory preparations used to generate them contained DNA in both the A- and B-forms. Having discovered the distinction between the crystalline and wet forms of DNA, she knew before anyone else that these previously accepted data were now based on qualitatively questionable laboratory preparations. She would therefore see the Pauling and Corey model as beset with serious conjectural problems regarding supporting evidence. See Judson, *Eighth Day,* 133, 157–59, 164.
69. Bazerman, "What Written Knowledge Does," 368.
70. Ibid.
71. All quotes are from *Double Helix,* 139. Judson explains the different forms bases can take; *Eighth Day,* 125–26.
72. For historical understanding of how these and other issues emerged and were resolved see Judson, *Eighth Day,* 113–75; Olby, *Path to the Double Helix,* 353–416; and also the concise discussion of Watson and Crick's collaboration up until publication of the model, in Allen, *Life Science,* 209–21.
73. For discussion of this point see Judson, *Eighth Day,* 166–167.
74. Furberg had published crystallographic results in 1949. His "standard configuration" established that individual nucleotides were three-dimensional, with the sugars at right angles to the bases (thus correcting Astbury's error, which put sugars parallel to bases). Furberg had entertained a helical model for DNA as early as 1949, but did not publish his views until late 1952 in *Acta Chemical Scandinavica.* See Judson, *Eighth Day,* 115.
75. See *Double Helix,* 130.
76. Bazerman, "What Written Knowledge Does," 366.
77. Halloran identifies explanatory power and what I am calling empirical adequacy as two of three main substantive arguments at work in the Watson and Crick article. The third line of appeal is elegance. I intend to show that the discourse makes explicit or implicit appeals to other lines of argument that went unnoticed by Halloran. See Halloran, "Birth of Molecular Biology," 73–74.
78. Quoted in Judson, *Eighth Day,* 143.

79. Erwin Chargaff, "Building the Tower of Babble," *Nature* 248 (1974): 778.
80. Judson, *Eighth Day,* 157.
81. Watson's private correspondence shows he was worried about the empirical adequacy of the model. He and Crick adduced consistency as a justification for accepting the model, at least tentatively, and inserted tentative language into their claims for adequacy. The experimental reports accompanying their article in *Nature* enabled Watson and Crick to add citation of some additional evidential support for their claim to empirical adequacy. See James D. Watson, letter to Max Delbrück, Mar. 12, 1953; in *Double Helix,* center insert.
82. Wilkins et al., "Molecular Structure," 739.
83. Franklin and Gosling, "Molecular Configuration," 741.
84. For discussion of specific empirical connections between these two experimental papers and the Watson and Crick model see Francis Crick, "The Double Helix: A Personal View," *Nature* 248 (1974): 766.
85. Judson, *Eighth Day,* 178.
86. Wilkins suggested they delete this sentence from a draft of their article, as it was too "ironical" now that much of that material would be published along with the bihelical model: "It is known that there is much unpublished experimental material." Quoted in *ibid.*
87. Judson, *Eighth Day,* 159–60; Watson, *Double Helix,* 106–07.
88. Crick discerned the dyadic symmetry of the DNA molecule from this important information. See Judson, *Eighth Day,* 164–67.
89. The question of whether Perutz acted ethically in giving Crick the report was raised by Andre Lwoff and Erwin Chargaff. Perutz subsequently published an *apologia,* which when taken together with the critical reviews provides excellent illustration of a conflict about what it means to think and to act responsibly as scientists. This dispute turned on whether the *topos* of communality could be used to warrant Perutz's action as situationally reasonable. Perutz argued that the report was not confidential. It was submitted to the Biophysics Committee which "served to exchange information" and, indeed, existed "to establish contact between the groups of people working for the Council in this field" (p. 1537). His action was therefore justified on the ground of communality.

    Others did not initially see the reasonableness of applying this *topos* in justifying Perutz's conduct, at least as circumstances were described in Watson's *Double Helix.* Lwoff saw the "gift" as a possible "breach of faith" (p. 134), and Chargaff scorned the violation of Franklin's right to secrecy (p. 1449). Watson later published a letter apologizing to Perutz for having created the wrong impression, but he also underscored how vital the report's information was to solving the DNA structure. See Erwin Chargaff, "A Quick Climb Up Mount Olympus," rev. of *The Double Helix,* by James D. Watson, *Science* 159 (1968): 1449; Andre Lwoff, "Truth, Truth, What Is Truth (About How the Structure of DNA Was Discovered)?" *Scientific American* 219 (1968): 134; Max F. Perutz, letter to the editor, *Science* 164 (1969): 1537–39; James D. Watson, letter to the editor, *Science* 164 (1969): 1539, and his *Double Helix,* 114–15.

90. *Double Helix,* 139.
91. Crick, "Double Helix: A Personal View," 766.
92. See J. D. Watson and F. H. C. Crick, "Genetical Implications of the Structure of Deoxyribonucleic Acid," *Nature* 171 (1953): 964–67.
93. Crick, "Double Helix: A Personal View," 766.
94. See Judson, *Eighth Day,* 184–93.
95. Max Delbrück pointed to these complex problems when he said of their structure: "In retrospect, what the dénouement was, was that both the principle of replication and the principle of readout are very simple, and the actual machinery for doing it is immensely complex. That's the way it has turned out" (quoted in Judson, *Eighth Day,* 60–61). Chargaff said of the second Watson and Crick paper, which explored these genetic implications: "I was much less in agreement with the scheme for DNA replication proposed in the second note; and even now, 20 years and thousands of experiments later, I cannot say that I am reconciled with it completely, the mechanism of DNA synthesis *in vivo* still being obscure to me"; "Building the Tower of Babble," 778. For other responses to the replication mechanism see Crick, "Double Helix: A Personal View," 768.
96. Halloran, "Birth of Molecular Biology," 73.
97. Chargaff, "Building the Tower of Babble," 778.
98. *Double Helix,* 131.
99. Crick, "Double Helix: A Personal View," 768 (emphasis added).
100. Discussion of x-ray crystallography is based on Allen, *Life Science,* 192–95, and Judson, *Eighth Day,* 76, 105–8, 121–24.
101. Discussion of chemical model-building is grounded in Allen, *Life Science* 171–72, 217, and Judson, *Eighth Day,* 83.
102. Judson, *Eighth Day,* 138.
103. Ibid., 128.
104. Ibid., 134–35, 141.
105. Quoted in ibid., 141.
106. Ibid., 146.
107. Rosalind E. Franklin and R. G. Gosling, "Evidence for the 2-Chain Helix in Crystalline Structure of Sodium Deoxyribonucleate," *Nature* 172 (1953): 156–57.
108. Judson, *Eighth Day,* 155.
109. Ibid., 151.
110. Quoted in ibid., 172. Also see Aaron Klug, "Rosalind Franklin and the Double Helix," *Nature* 248 (1974): 787–88.
111. "Molecular Configuration," 740.
112. Ibid.
113. Ibid., 741.
114. Ibid.
115. Quoted in Judson, *Eighth Day,* 113–14. Crick also asserted elsewhere that both King's College groups failed to solve the base-pairing problem because "they had done no proper model building"; "Double Helix: A Personal View," 766.
116. "Building the Tower of Babble," 778.

CHAPTER 11

1. R. Michael Bokeno, "The Rhetorical Understanding of Science: An Explication and Critical Commentary," *Southern Speech Communication Journal* 52 (1987): 311.
2. For example, see Richard M. Weaver, "Concealed Rhetoric in Scientistic Sociology," in Richard L. Johannesen, Rennard Strickland, and Ralph T. Eubanks, eds., *Language Is Sermonic: Richard M. Weaver and the Nature of Rhetoric* (Baton Rouge: Louisiana State University Press, 1970), 139–58; Weaver, "The Rhetoric of Social Science," in his *The Ethics of Rhetoric* (1953; rpt. Davis, CA: Hermagoras Press, 1985), 186–210; and Andrew Weigert, "The Immoral Rhetoric of Scientific Sociology," *The American Sociologist* 5 (1970): 111–19.
3. See Perelman and Olbrechts-Tyteca, *The New Rhetoric,* (Notre Dame, IN: University of Notre Dame Press, 1969); Grassi, *Rhetoric as Philosophy: The Humanist Tradition* (University Park: Pennsylvania State University Press, 1980); Weimer, "Science as a Rhetorical Transaction: Toward a Nonjustificational Conception of Rhetoric," *Philosophy and Rhetoric* 10 (1977): 1–29. Fisher, *Human Communication as Narration* (Columbia: University of South Carolina Press, 1987); Apostel, "Assertion Logic and Theory of Argumentation," *Philosophy and Rhetoric* 4 (1971): 92–110; and Toulmin, *The Uses of Argument* (Cambridge: Cambridge University Press, 1958).
4. For instance, Paul Feyerabend, "Democracy, Elitism, and Scientific Method," *Inquiry* 23 (1980): 3–18; and Philip C. Wander, "The Rhetoric of Science," *Western Speech Communication Journal* 40 (1976): 227–29.
5. A preliminary list of themes for making political arguments is presented in Michael Calvin McGee and John R. Lyne, "What Are Nice Folks like You Doing in a Place like This?" in John S. Nelson et al., eds., *The Rhetoric of the Human Sciences* (Madison: University of Wisconsin Press, 1987), 399–400.
6. While essays in *Rhetoric of the Human Sciences* imply at varying levels that we must attend to inventional processes for full understanding of the rhetoric of the human sciences, those essays do not make inventional processes the centerpiece of rhetorical investigations. I hope to have offered a way of understanding and analyzing inventional practices that can be useful to these authors' specific concern with the "rhetoric of inquiry." General theory of rhetorical invention can be adjusted heuristically to articulate field-variant, special inventional systems that account for what it means to practice the "rhetoric of" anthropology, psychology, economics, political science, history, theology, law, and other fields of inquiry. One advantage is that commentators would have a common rhetorical language, grounded in inventional precepts, which cuts across the technical considerations of any area chosen for exploration. Another advantage is that flexible adaptation of general theory to the special problems, norms, and expectations of particular fields will yield field-specific frameworks, collectively allowing scrutiny of similarities and differences among the rhetorics of several fields. Finally, though some essayists in the Nelson et al. volume touch on the role of *topoi* and other inventional heuristics, the possibility of

these constituting the features of an organized, informal, logical system for a discipline is not spelled out there.

Other useful statements of central presuppositions supporting the rhetoric of inquiry project can be found in John S. Nelson and Allan Megill, "Rhetoric of Inquiry: Projects and Prospects," *Quarterly Journal of Speech* 72 (1986): 20–37; Herbert W. Simons, "Chronicle and Critique of a Conference," *Quarterly Journal of Speech* 71 (1985): 52–64; and John Lyne, "Rhetorics of Inquiry," *Quarterly Journal of Speech* 71 (1985): 65–73.

NAME INDEX

Name Index

SUBJECT INDEX

BARTON COLLEGE LIBRARY
500 P915r
nonf
Prelli, Lawrence J./A rhetoric of scienc
3 6500 00052 3974